STUDENT SOLUTIONS MANUAL FOR KASEBERG'S
INTRODUCTORY ALGEBRA
A JUST-IN-TIME APPROACH

Cindy Ann Rubash
and
John Thickett
Southern Oregon State College

D1737414

Brooks/Cole Publishing Company

I(T)P® An International Thomson Publishing Company

Pacific Grove • Albany • Belmont • Bonn • Boston • Cincinnati • Detroit • Johannesburg • London
Madrid • Melbourne• Mexico City • New York • Paris • Singapore • Tokyo •Toronto • Washington

Assistant Editor: *Beth Wilbur*
Cover Design: *Vernon T. Boes*
Cover Illustration: *Harry Briggs*
Editorial Associate: *Nancy Conti*
Marketing Team: *Maureen Riopelle and Deborah Petit*
Production Editor: *Mary Vezilich*
Printing and Binding: *Patterson Printing*

 The ITP logo is a registered trademark under license.

For more information, contact:

BROOKS/COLE PUBLISHING
511 Forest Lodge Road
Pacific Grove, CA 93950
USA

International Thomson Editores
Seneca 53
Col. Polanco
11560 México, D. F., México

International Thomson Publishing Europe
Berkshire House 168-173
High Holborn
London WC1V 7AA
England

International Thomson Publishing Japan
Hirakawacho Kyowa Building, 3F 418
2-2-1 Hirakawacho
Chiyoda-ku, Tokyo 102
Japan

Thomas Nelson Australia
102 Dodds Street
South Melbourne, 3205
Victoria, Australia

International Thomson Publishing Asia
221 Henderson Road
#05-10 Henderson Building
Singapore 0315

Nelson Canada
1120 Birchmount Road
Scarborough, Ontario
Canada M1K 5G4

International Thomson Publishing GmbH
Königswinterer Strasse
53227 Bonn
Germany

Printed in the United States of America

5   4   3

ISBN 0-534-95668-8

CONTENTS

Section 1.0

1. a) Product
   b) Variable
   c) Expression
   d) Square of x
   e) Sum

3. a) Opposites
   b) Factors
   c) Reciprocals
   d) Factoring
   e) Sets

5. a) $-15 + (-3) = -18$

   $-15 - (-3) = -15 + 3 = -12$

   $-15 \cdot (-3) = 45$

   $-15 \div (-3) = 5$

   b) $-6 + 2 = -4$

   $-6 - 2 = -6 + (-2) = -8$

   $-6 \cdot 2 = -12$

   $-6 \div 2 = -3$

   c)

   $$\frac{1}{4} + \left(-\frac{1}{2}\right) = \frac{1}{4} - \frac{1}{2} = \frac{1}{4} - \left(\frac{1}{2} \cdot \frac{2}{2}\right) = \frac{1}{4} - \frac{2}{4} = -\frac{1}{4}$$

   $$\frac{1}{4} - \left(-\frac{1}{2}\right) = \frac{1}{4} + \frac{1}{2} = \frac{1}{4} + \frac{2}{4} = \frac{3}{4}$$

   $$\frac{1}{4} \cdot \left(-\frac{1}{2}\right) = \frac{1}{4} \cdot \frac{(-1)}{2} = -\frac{1}{8}$$

   $$\frac{1}{4} \div \left(-\frac{1}{2}\right) = \frac{1}{4} \cdot \left(-\frac{2}{1}\right) = \frac{1 \cdot (-2)}{4 \cdot 1} = -\frac{2}{4} = -\frac{1}{2}$$

   d)

   $$-\frac{5}{12} + \left(-\frac{1}{3}\right) = -\frac{5}{12} + \left(-\frac{1}{3} \cdot \frac{4}{4}\right)$$

   $$= -\frac{5}{12} + \left(-\frac{4}{12}\right) = -\frac{9}{12} = -\frac{3}{4}$$

   $$-\frac{5}{12} - \left(-\frac{1}{3}\right) = -\frac{5}{12} + \frac{1}{3} = -\frac{5}{12} + \frac{4}{12} = -\frac{1}{12}$$

   $$-\frac{5}{12} \cdot \left(-\frac{1}{3}\right) = \left(\frac{-5 \cdot -1}{12 \cdot 3}\right) = \frac{5}{36}$$

   $$-\frac{5}{12} \div \left(-\frac{1}{3}\right) = -\frac{5}{12} \cdot \left(-\frac{3}{1}\right)$$

   $$= \frac{-5 \cdot -3}{12 \cdot 1} = \frac{15}{12} = \frac{5}{4}$$

5. e) $2.5 + (-0.25) = 2.50 - 0.25 = 2.25$

   $2.5 - (-0.25) = 2.50 + 0.25 = 2.75$

   $2.5 \cdot (-0.25) = -0.625$

   $2.5 \div (-0.25) = -10$

7. a) $4 + 2 + 8 + 6 = 4 + 6 + 2 + 8$
         (Commutative)
   $= (4 + 6) + (2 + 8) = 10 + 10 = 20$
   (Associative)

   b) $5 \cdot 7 \cdot 2 \cdot 3 = 5 \cdot 2 \cdot 7 \cdot 3$
         (Commutative)
   $= (5 \cdot 2) \cdot (7 \cdot 3) = 10 \cdot 21 = 210$
   (Associative)

   c) $5 + -2 + 15 + 12 = 5 + 15 + 12 + (-2)$
         (commutative)
   $= (5 + 15) + (12 - 2)$
   (Associative)
   $= 20 + 10 = 30$

9. a) $-2(3 - 4x) = -2 \cdot 3 + (-2)(-4x)$
         (Disributive)
   $= -6 + (-2 \cdot (-4))x = -6 + 8x$
   (Associative)

   b) $-a(b - c) = -a \cdot b + (-a)(-c)$
         (Distributive)
   $= -ab + ac$

11. a) $6x - 54 = 6 \cdot x + 6 \cdot (-9)$

   $= 6(x + (-9)) = 6(x - 9)$

   b) $15x - 225 = 15 \cdot x + 15 \cdot (-15)$

   $= 15(x + (-15) = 15(x - 15)$

13. a) $3a - 2b + 4c - 2a$

    $= 3a - 2a - 2b + 4c$

    $= a - 2b + 4c$

   b) $bx^2 + 2x - 4x^2 - 3$

    $= 6x^2 - 4x^2 + 2x - 3$

    $= 2x^2 + 2x - 3$

15. a)

Multiply	x	-2
3x	$3x \cdot x =$ $3x^2$	$3x \cdot (-2) =$ $-6x$

   b)

Factor	a	-b
2a	$2a^2 =$ $2a \cdot a$	$-2ab =$ $2a \cdot (-b)$

17. a)

	2x+4 mm
3 mm	$3(2x+4)$ (mm)(mm) = $3 \cdot 2x + 3 \cdot 4$ mm^2 $= 6x + 12mm^2$

   b)

	3y+2 km
5 km	$15y+10$ km^2 = $5 \cdot 3y + 5 \cdot 2$ $km \cdot km$

19. $3 \cdot \boxed{x^2}$ $5 \boxed{x}$ $6 \boxed{ 1}$ $= 3x^2 + 5x + 6$.

21. a) J
    b) E
    c) G
    d) K

23. a) $\approx (-17, 1)$
    b) A(-40, -40)

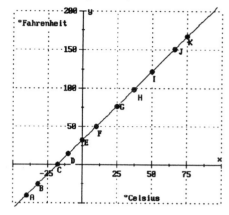

25.

x	$y = 2x + 3$
-3	$2(-3) + 3 = -6 + 3 = -3$
-2	$2(-2) + 3 = -4 + 3 = -1$
-1	$2(-1) + 3 = -2 + 3 = 1$
0	$2(0) + 3 = 0 + 3 = 3$
1	$2(1) + 3 = 2 + 3 = 5$
2	$2(2) + 3 = 4 + 3 = 7$
3	$2(3) + 3 = 6 + 3 = 9$

27.

x	$y = x^2 - x$
-3	$(-3)^2 - (-3) = 9 + 3 = 12$
-2	$(-2)^2 - (-2) = 4 + 2 = 6$
-1	$(-1)^2 - (-1) = 1 + 1 = 2$
0	$(0)^2 - (0) = 0 - 0 = 0$
1	$(1)^2 - (1) = 1 - 1 = 0$
2	$(2)^2 - (2) = 4 - 2 = 2$
3	$(3)^2 - (3) = 9 - 3 = 6$

29.

x	$y = 2 - x^2$
-3	$2 - (-3)^2 = 2 - 9 = -7$
-2	$2 - (-2)^2 = 2 - 4 = -2$
-1	$2 - (-1)^2 = 2 - 1 = 1$
0	$2 - (0)^2 = 2 - 0 = 2$
1	$2 - (1)^2 = 2 - 1 = 1$
2	$2 - (2)^2 = 2 - 4 = -2$
3	$2 - (3)^2 = 2 - 9 = -7$

31.

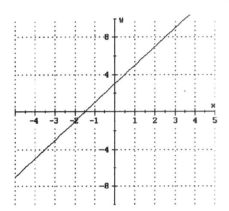

33.

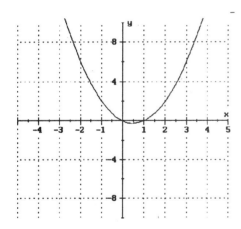

35.

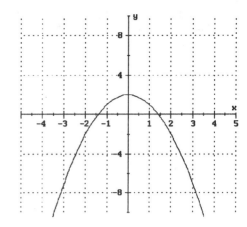

37. b) - A product to a sum

39. e) - $x \geq 0$

41. a) - Subtraction to addition of the opposite number.

43. a) Distributive property does not apply - can not distribute multiplication over multiplication

b) May be changed
c) May be changed

45. Subtraction may be changed to addition of the opposite number.

47. The product of numbers with like signs is positive.

3

49. a)

Input x	-1	8	5	-3	-4	5
Input y	6	-3	-2	6	7	0
Input x+y	5	5	3	3	3	5

b)

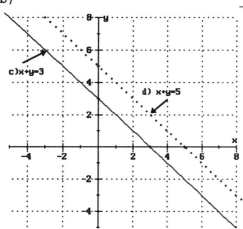

e) Answers will vary.

51. $p - x \leq 2$

$t \cdot y > 2$

53. a) $-2 < x < 4$

b) $x \leq 1$ or $x \geq 3$

c) $x < 2$ or $x > 4$

d) $-1 < x < 4$

Section 1.1

1. a)

x	$y = 15x - 4$
1	$15(1) - 4 = 15 - 4 = 11$
2	$15(2) - 4 = 30 - 4 = 26$
3	$15(3) - 4 = 45 - 4 = 41$
4	$15(4) - 4 = 60 - 4 = 56$

b) $f(x) = 15x - 4$

3. a)

h	$d = \sqrt{\dfrac{3h}{2}}$
5	$\sqrt{\dfrac{3(5)}{2}} = \sqrt{\dfrac{15}{2}} \approx 2.739$
10	$\sqrt{\dfrac{3(10)}{2}} = \sqrt{\dfrac{30}{2}} \approx 3.873$
15	$\sqrt{\dfrac{3(15)}{2}} = \sqrt{\dfrac{45}{2}} \approx 4.743$
20	$\sqrt{\dfrac{3(20)}{2}} = \sqrt{\dfrac{60}{2}} \approx 5.477$

b) $f(x) = \sqrt{\dfrac{3h}{2}}$

5. a)

y	$x = y^2 + 2$
-4	$(-4)^2 + 2 = 16 + 2 = 18$
-2	$(-2)^2 + 2 = 4 + 2 = 6$
0	$(0)^2 + 2 = 0 + 2 = 2$
2	$(2)^2 + 2 = 4 + 2 = 6$
4	$(4)^2 + 2 = 16 + 2 = 18$

b) Not a function.

7. a) 6:00 am
   b) Undefined
   c) 6:00 am
   d) 7:00 am

9. a) $f(1) = 3 + 2[(1) - 1] = 3 + 2(0) = 3 + 0 = 3$

   b) $f(5) = 3 + 2[(5) - 1] = 3 + 2(4) = 3 + 8 + 11$

   c) $f(n) = 3 + 2[(n) - 1] = 3 + 2n - 2$
   $$= 2n + 3 - 2 = 2n + 1$$

   d) $f(n + m) = 3 + 2[(n + m) - 1]$
   $$= 3 + 2(n + m) - 2$$
   $$= 2n + 2m + 3 - 2$$
   $$= 2n + 2m + 1$$

11. a) $f(3) = (3)^2 + (3) - 2 = 9 + 3 - 2 = 10$

    b) $f(1) = (1)^2 + (1) - 2 = 1 + 1 - 2 = 0$

    c) $f(\square) = (\square)^2 + (\square) - 2 = \square^2 + \square - 2$

    d) $f(n) = (n)^2 + (n) - 2 = n^2 + n - 2$

    e) $f(n - m) = (n - m)^2 + (n - m) - 2$
    $$= n^2 - 2nm + m^2 + n - m - 2$$

13. a) from table 17 {-2, 4}
    b) from table 17 {-1, 3}
    c) extending table 17 {-4, 6}
    $$f(-3) = (-3)^2 - 2(-3) - 3 = 12$$
    $$f(-4) = (-4)^2 - 2(-4) - 3 = 21$$
    $$f(5) = (5)^2 - 2(5) - 3 = 12$$
    $$f(6) = (6)^2 - 2(6) - 3 = 21$$

    d) extending table 17 {-7, 9}
    $$f(-5) = (-5)^2 - 2(-5) - 3 = 32$$
    $$f(-6) = (-6)^2 - 2(-6) - 3 = 45$$
    $$f(-7) = (-7)^2 - 2(-7) - 3 = 60$$
    $$\left.\begin{array}{l} f(7) = 32 \\ f(8) = 45 \\ f(9) = 60 \end{array}\right\} \text{from symmetry}$$

    e) expanding table 17 ≈{-1.83, 3.83}
    $$f(-1.5) = (-1.5)^2 - 2(-1.5) - 3 = 2.25$$
    $$f(-1.75) = (-1.75)^2 - 2(-1.75) - 3 = 3.5625$$
    $$f(-1.8) = (-1.8)^2 - 2(-1.8) - 3 = 3.84$$
    $$f(-1.83) = (-1.83)^2 - 2(-1.83) - 3 = 4.0089$$
    $$f(-3.83) = 4.0089 \text{ due to symmetry.}$$
    actual solution $\left\{1 - 2\sqrt{2}, 1 + 2\sqrt{2}\right\}$

15. a) $y = \sqrt{x}$ is the upper
    half of $x = y^2$

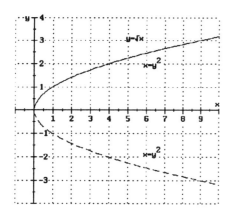

b) Answers may vary
   Possible solutions are;

   $\sqrt{4} = 2$ and $4 = 2^2$

   $\sqrt{9} = 3$ and $9 = 3^2$

   $\sqrt{16} = 4$ and $16 = 4^2$

c) $y = \sqrt{x}$ restricts y to $y \geq 0$

   $x = y^2$ has no restriction on y

d) in $y = \sqrt{x}$, x must be ≥ 0

   in $x = y^2$, x will always be ≥ 0

17. From the table locate desired
    output and corresponding input.
    From the graph locate output on
    y-axis, trace horizontally to
    graph, trace vertically to x-
    axis for input.

    a) $\{-5,5\}$
    b) $\{0\}$
    c) $\{ \}$ (empty set-no solution)

19. a) $f(r) = 2\pi r, f(r) = C$
    b) $f(r) = \pi r^2, f(r) = A$

21. a) $f(l,w,h) = l \cdot w \cdot h, f(l,w,h) = V$
    b) $f(r,h) = \pi r^2 h, f(r,h) = V$

23. $f(P,r) = P \cdot r, f(P,r) = I$

25. a)

x	$f(x) = \lvert x - 2 \rvert$
-3	$\lvert(-3) - 2\rvert = \lvert-5\rvert = 5$
-1	$\lvert(-1) - 2\rvert = \lvert-3\rvert = 3$
0	$\lvert(0) - 2\rvert = \lvert-2\rvert = 2$
1	$\lvert(1) - 2\rvert = \lvert-1\rvert = 1$
3	$\lvert(3) - 2\rvert = \lvert1\rvert = 1$
5	$\lvert(5) - 2\rvert = \lvert3\rvert = 3$

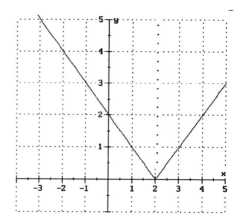

b) Answers may vary;
   f(3) = f(1) = 1
   f(-1) = f(5) = 3

c) x = 2

27. Fails vertical line test,
    not a function
    Line of symmetry y = 0

29. Passes vertical line test,
    is a function
    Line of symmetry x = 3

31. a) $x \geq 0$
    b) $x - 2 \geq 0$; $x \geq 2$
    c) $x + 3 \geq 0$; $x \geq -3$

33. a) $x \geq 3$; $x - 3 \geq 0, f(x) = \sqrt{x-3}$
    b) $x \geq -4$; $x + 4 \geq 0, f(x) = \sqrt{x+4}$
    c) $x \geq -1$; $x + 1 \geq 0, f(x) = \sqrt{x+1}$

35. Graphs move left and down

Section 1.1

37. 10 numbers.

39. 9 numbers

41. Exercise 28

43. Exercise 29

45. Refer to graph in exercise 25.
     Domain: $x = \mathbb{R}$
     Range: $y \geq 0$

47. Domain: $x = \mathbb{R}$
     Range: $y = \mathbb{R}$

Section 1.2

1.

Δx	x	y	Δy
10 <	10	0.60	> 0.6
10 <	20	1.20	> 0.6
10 <	30	1.80	> 0.6
	40	2.40	

Linear function.

$slope = \dfrac{0.60}{10} = \$0.06 \ tax \ per \ \$ \ sales$

3.

Δx	x	y	Δy
1 <	0	20.00	> −0.75
1 <	1	19.25	> −0.75
1 <	2	18.50	> −0.75
	3	17.75	

Linear function.

$slope = \dfrac{-0.75}{1} = -\$0.75 \ per \ trip$

5.

Δx	x	y	Δy
.5 <	0	6.0	> 46
.5 <	0.5	46.0	> 31.9
.5 <	1.0	77.9	> 23.9
.5 <	1.5	101.8	> 15.8
	2.0	117.6	

Non-linear.

7. (1,3) to (3,4) $slope = \dfrac{4-3}{3-1} = \dfrac{1}{2}$

(1, 3) to (2, 1) $slope = \dfrac{1-3}{2-1} = -2$

(2, 1) to (4, 2) $slope = \dfrac{2-1}{4-2} = \dfrac{1}{2}$

(3, 4) to (4, 2) $slope = \dfrac{2-4}{4-3} = -2$

$\dfrac{1}{2} \cdot (-2) = -1$

Opposite sides are parallel.
Adjacent sides are perpendicular.

9.

a) (2, 5) *to* (4, 1); slope $= \dfrac{1-5}{4-2} = \dfrac{-4}{2} = -2$

b) (1, 2) *to* (5, 4); slope $= \dfrac{4-2}{5-1} = \dfrac{2}{4} = \dfrac{1}{2}$

$-2 \cdot \dfrac{1}{2} = -1$ diagonals are perpendicular

11. a) (5,0) and (0,−3); slope $= \dfrac{-3-0}{0-5} = \dfrac{-3}{-5} = \dfrac{3}{5}$

b) (0,4) and (−5,0); slope $= \dfrac{0-4}{-5-0} = \dfrac{-4}{-5} = \dfrac{4}{5}$

13. Slope is the opposite of the y-intercept divided by the x-intercept.

15.

$\left(0, \ 6\dfrac{1}{3}\right)$ to $\left(9\dfrac{1}{2}, \ 0\right)$

$= \left(0, \ \dfrac{19}{3}\right)$ to $\left(\dfrac{19}{2}, \ 0\right)$

$slope = \dfrac{0 - \dfrac{19}{3}}{\dfrac{19}{2} - 0}$

$= -\left(\dfrac{\dfrac{19}{3}}{\dfrac{19}{2}}\right) = -\left(\dfrac{19}{3} \cdot \dfrac{2}{19}\right)$

$= -\dfrac{2}{3}$

17. a) 0 (flat line)
b) Positive (goes up)
c) 0 (flat line)
d) Day 8 to 9 (steeper)
e) 2 (day 2 to 3 & day 8 to 9)

Section 1.2

19. a)

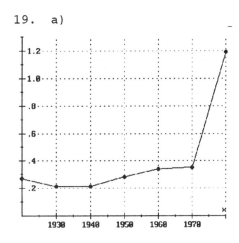

b) (1920,0.27) to (1930,0.21)

$$\text{Slope} = \frac{0.21 - 0.27}{1930 - 1920} = \frac{-.06}{10} = -0.006$$

(1930,0.21) to (1940,0.21)

$$\text{Slope} = \frac{0.21 - 0.21}{1940 - 1930} = \frac{0}{10} = 0$$

(1940,0.21) to (1950,0.28)

$$\text{Slope} = \frac{0.28 - 0.21}{1950 - 1940} = \frac{0.07}{10} = 0.007$$

(1950,0.28) to (1960,0.34)

$$\text{Slope} = \frac{0.34 - 0.28}{1960 - 1950} = \frac{0.06}{10} = 0.006$$

(1960,0.34) to (1970,0.35)

$$\text{Slope} = \frac{0.35 - 0.34}{1970 - 1960} = \frac{0.01}{10} = 0.001$$

(1970,0.35) to (1980,1.19)

$$\text{Slope} = \frac{1.19 - 0.35}{1980 - 1970} = \frac{0.84}{10} = 0.084$$

c) 1970 to 1980

d) Dollars per year,
   Change in price per year.

21.

Δx	Temp, x (°C)	Wind Chill, y (°C)	Δy	Slope
	20	13		$\frac{-18}{-12} = \frac{3}{2}$
-12			-18	
	8	-5		$\frac{-12}{-8} = \frac{3}{2}$
-8			-12	
	0	-17		$\frac{-6}{-4} = \frac{3}{2}$
-4			-6	
	-4	-23		$\frac{-18}{-12} = \frac{3}{2}$
-12			-18	
	-16	-41		$\frac{-6}{-4} = \frac{3}{2}$
-4			-6	
	-20	-47		

Linear relationship

23. False, zero is in the denominator

25. True

27. False, means undefined slope

29. a) {-1.5,4.5}
     b) {-3.5,6.5}
     c) {}; no solution

31. y-intercept is 3.
     Input is x = 0

33. x-intercepts -500 & -300
     They have no meaning,
     cannot be negative miles.

35. (0,0), no sale, no tax.

37. $\left(26\frac{2}{3}, 0\right)$ # of trips;

     (0, $20) initial value of ticket.
     20 - 0.75x = 0, 20 = 0.75x

     $x = 26\frac{2}{3}$

39. x<0 & y>0

41. x<0 & y>0
         or
     x>0 & y<0

43. Quadrant 2

45. Quadrant 3

47. Quadrants 3&4

49. Quadrants 2&3

57.

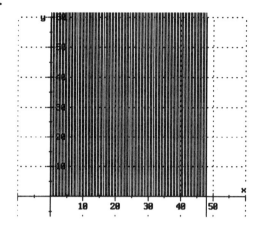

51.

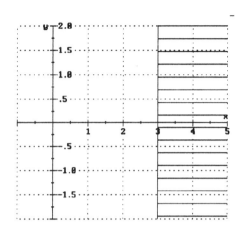

59.

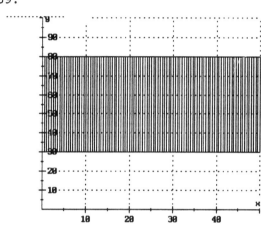

53.

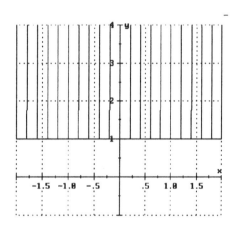

61. Answers will vary. One possible solution:

55.

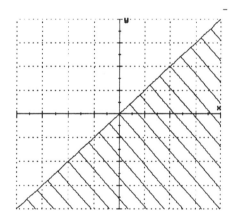

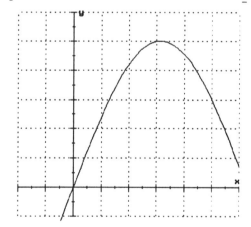

63. a)

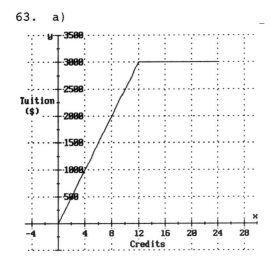

b)

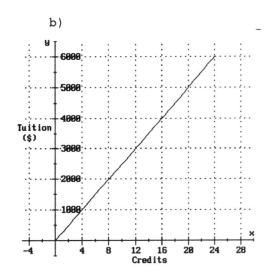

65.

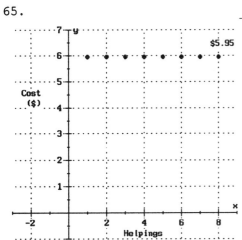

67.

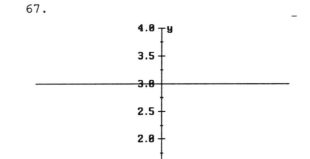

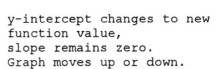

y-intercept changes to new
function value,
slope remains zero.
Graph moves up or down.

Mid Chapter 1 Test

1. Quadrants 2&3; (x<0)

2. (x,y) or x&y coordinates

3. Associative

4. y-axis, vertical axis.

5. Vertical line text

6. Symmetry

7. Domain

8. Constant

9. a) $f(1) = (1)^2 - (1) = 1 - 1 = 0$
    b) $f(3) = (3)^2 - (3) = 9 - 3 = 6$
    c) $f(-2) = (-2)^2 - (-2) = 4 + 2 = 6$
    d) $f(a) = (a)^2 - (a) = a^2 - a$
    e) $f(a + b) = (a + b)^2 - (a + b)$
       $= a^2 + 2ab + b^2 - a - b$

10. a) $y > 0$
    b) $x < 0$
    c) $y < 0$
    d) $x \geq 0$
    e) $x > 0$
    f) $y \leq 0$

11. a) {-1,5}
    b) {1,3}
    c) {2}

12. Matching outputs are equi-
    distant from the line of
    symmetry, x = 2.

13. a)

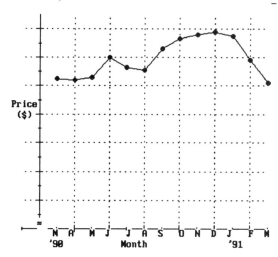

b) (Aug,1.11) to (Sept,1.26);
   slope $= \dfrac{1.26 - 1.11}{9 - 8} = \dfrac{0.15}{1} = 0.15$

c) (Mar,1.05) to (Apr,1.04);
   slope $= \dfrac{1.04 - 1.05}{4 - 3} = \dfrac{-0.01}{1} = -0.01$

d) (Jan,1.35) to (Feb,1.18)
   slope $= \dfrac{1.18 - 1.35}{2 - 1} = \dfrac{-0.17}{1} = -0.17$

e) Jan to Feb
f) Change in price per month.

14. T = tax, S = Dollars spent;
    T = 0.07S

15. y = Cost, x = Purchase in
    dollars; $y = x+.07x = 1.07x$

16-21:
    a) 16, 17, 19, 21
    b) 18, 20
    c) 18, 20
    d) 16,21

Section 1.3

1.

Side 1	Side 2	Side 3	Total
1	3 + 6(1) = 9	7(1) − 3 = 4	14
5	3 + 6(5) = 33	7(5) − 3 = 32	70
6	3 + 6(6) = 39	7(6) − 3 = 39	84
7	3 + 6(7) = 45	7(7) − 3 = 46	98
⇒ 6.5	3 + 6(6.5) = 42	7(6.5) − 3 = 42.5	91
x	3 + 6x	7x − 3	14x

3.

Meal	Tip	Tax	Total
$10.00	0.15($10.00) = $1.50	$1.50 ÷ 25 = $0.40	$11.90
$20.00	0.15($20.00) = $3.00	$3.00 ÷ 25 = $1.20	$24.20
$30.00	0.15($30.00) = $4.50	$4.50 ÷ 25 = $1.80	$36.30
$25.00	0.15($25.00) = $3.75	$3.75 ÷ 25 = $1.50	$30.25
⇒ $28.00	0.15($28.00) = $4.20	$4.20 ÷ 25 = $1.68	$33.88
x	0.15x	0.06x	1.21x

5. Slope = 0.055; y-intercept = 0

7. Slope = $3.00; y-intercept = $10

9.
$y = \$85 + \$4.75(x - 10)$, $x > 10$
$y = \$85 + \$4.75x - \$47.50$
$y = \$4.75x + \$37.50$
Slope = $4.75; y − intercept = $37.50
Inequality makes y − intercept meaningless

11.
$y = \$0.26 + \$0.19(x - 1)$, $x \geq 1$
$y = \$0.26 + \$0.19x - \$0.19$,
$y = \$0.19x + \$0.07$
Slope = $0.19; y − intercept = $0.07

13. Slope = 2π

15. Slope = μ

17. Slope = b

19. $\$2.16 = \$0.19(11) + b$;
$\$2.16 - \$2.09 = b$,
$b = \$0.07$

21. $y = \$0.26 + \$0.19(x - 1)$, $x \geq 1$
$y = \$0.26 + \$0.19x - \$0.19$,
$y = \$0.19x + \$0.07$
(NOTE: Inputs are rounded to next whole number. 1½ additional minutes cost the same as 2 additional minutes.)

$f(2) = \$0.19(2) + \$0.07 = \$0.45$
$f(4) = \$0.19(4) + \$0.07 = \$0.83$
$f(4.5) = \$0.19(5) + \$0.07 = \$1.02$
$f(8) = \$0.19(8) + \$0.07 = \$1.59$
$f(8.5) = \$0.19(9) + \$0.07 = \$1.78$
$f(10.5) = \$0.19(11) + \$0.07 = \$2.16$

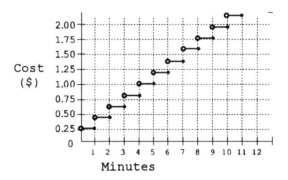

23. $m = \dfrac{4500 - 855}{250 - 60} = \dfrac{3645}{190} = \dfrac{729}{38}$
≈ 19.184
$b = y - mx = 855 - 19.184(60)$
$b = -296.04$
$y \approx 19.184x - 296$;
lines are nearly the same.

25. Fixed Cost = $300,
Variable Cost = 0.025 per dollar.
C = 0.025x + $300

27. Input = # Shoes;
(250,$11000) (300,$13185)
$m = \dfrac{13185 - 11000}{300 - 250} = \dfrac{2185}{50} = \$43.70$
$b = 11000 - 43.70(250) = \$75.00$
Fixed Costs = $75.00,
Variable Costs = $43.70 per pair.
C = $43.70x + $75.00

29. C = $5.95; Constant function.

Section 1.3

31. C = $5.95x; Increasing function.

33. C = $35; Constant function.

35. V = $35-$1.50x; Decreasing function.

37. Graph would rise faster.

39. Graph would move up to new y-intercept = $40.

41. a) Pulse rate is a function of age.
    b) Answers will vary.
    c) P = 0.5(220-x);
       where x = age.
    d) P = 0.7(220-x)
    e) $P_{(low)}$ = 0.5(220-50)
       = 0.5(170) = 85.
       $P_{(high)}$ = 0.7(220-50)
       =0.7(170) = 119
    f) 95 = 0.5(220 - x); 95 = 110 - 0.5x;

       0.5x = 15; x = 30

       133 = 0.7(220 - x); 133 = 154 - 0.7x;

       0.7x = 21; x = 30

43. a) from graph, $m = \dfrac{-3}{1.5} = -2$; b = 3

    $f(x) = -2x + 3$ for $x \le 1.5$

    b) $f(1.5) = -2(1.5) + 3 = -3 + 3 = 0$

    c) from graph, $m = \dfrac{3}{1.5} = 2$

       $b = 3 - 2(3) = 3 - 6 = -3$

       $f(x) = 2x - 3$ for $x \ge 1.5$

    d) $f(1.5) = 2(1.5) - 3 = 3 - 3 = 0$

45. Using Linear Regression
    $y \approx 10.1x - 13.8$

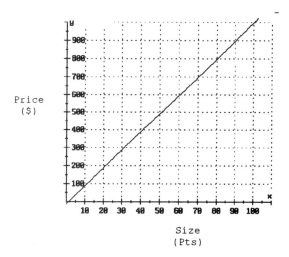

Price
($)

Size
(Pts)

47. Using Linear Regression
    a) $y \approx -45.2x - 11244.4$

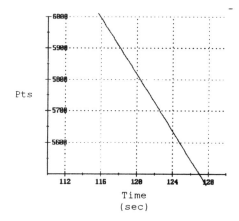

Pts

Time
(sec)

    b) $\dfrac{5600}{125} = 44.8$; $\dfrac{5705}{122} = 46.762$;

       $\dfrac{5915}{118} = 50.127$; $\dfrac{5950}{117} = 50.855$

       Points per unit time increase as time decreases.

Section 1.4

1. Take off jacket,
   take off vest,
   take off shirt;
   dressing and undressing.

3. Inverse is not meaningful;
   cannot 'untake' a picture.

5. Order not important.

7.
$$2x - 4 = 8$$
$$2x - 4 + 4 = 8 + 4$$
$$2x = 12$$
$$\frac{2x}{2} = \frac{12}{2}$$
$$x = 6$$

9.
$$2x - 4 = -6$$
$$2x - 4 + 4 = -6 + 4$$
$$2x = -2$$
$$\frac{2x}{2} = \frac{-2}{2}$$
$$x = -1$$

11.
$$\frac{1}{2}x + 1 = 5$$
$$\frac{1}{2}x + 1 - 1 = 5 - 1$$
$$\frac{1}{2}x = 4$$
$$\frac{2}{1} \cdot \frac{1}{2}x = 4 \cdot \frac{2}{1}$$
$$x = 8$$

13.
$$\frac{1}{2}x + 1 = 0$$
$$\frac{1}{2}x + 1 - 1 = 0 - 1$$
$$\frac{1}{2}x = -1$$
$$\frac{2}{1} \cdot \frac{1}{2}x = -1 \cdot \frac{2}{1}$$
$$x = -2$$

15.
$$3x + 3 = 0$$
$$3x + 3 - 3 = 0 - 3$$
$$3x = -3$$
$$\frac{3x}{3} = \frac{-3}{3}$$
$$x = -1$$

17.
$$\frac{1}{3}x - 1 = 4$$
$$\frac{1}{3}x - 1 + 1 = 4 + 1$$
$$\frac{1}{3}x = 5$$
$$\frac{3}{1} \cdot \frac{1}{3}x = 5 \cdot \frac{3}{1}$$
$$x = 15$$

19.
$$\frac{5}{9}(x - 32) = -25$$
$$\frac{9}{5} \cdot \frac{5}{9}(x - 32) = -25 \cdot \frac{9}{5}$$
$$x - 32 = \frac{-25 \cdot 9}{5} = -45$$
$$x - 32 + 32 = -45 + 32$$
$$x = -13$$

21.
$$\frac{5}{9}(x - 32) = 5$$
$$\frac{9}{5} \cdot \frac{5}{9}(x - 32) = 5 \cdot \frac{9}{5}$$
$$x - 32 = 9$$
$$x - 32 + 32 = 9 + 32$$
$$x = 41$$

23.
$$x - 2(4 - x) = -17$$
$$x - 8 + 2x = -17$$
$$3x - 8 = -17$$
$$3x - 8 + 8 = -17 + 8$$
$$3x = -9$$
$$\frac{3x}{3} = \frac{-9}{3}$$
$$x = -3$$

Section 1.4

25.
$$x - 3(4 + x) = 6$$
$$x - 12 - 3x = 6$$
$$-2x - 12 = 6$$
$$-2x - 12 + 12 = 6 + 12$$
$$-2x = 18$$
$$\frac{-2x}{-2} = \frac{18}{-2}$$
$$x = -9$$

27.
$$x + 3(5 - x) = -9$$
$$x + 15 - 3x = -9$$
$$-2x + 15 = -9$$
$$-2x + 15 - 15 = -9 - 15$$
$$-2x = -24$$
$$\frac{-2x}{-2} = \frac{-24}{-2}$$
$$x = 12$$

29.
$$50.40 = x + 0.05x;$$
$$50.40 = 1.05x$$
$$\frac{50.40}{1.05} = \frac{1.05x}{1.05}$$
$$48 = x \text{ or}$$
$$x = 48$$

31.
$$D = \overset{\downarrow}{r} t$$
$$\frac{D}{t} = \frac{rt}{t}$$
$$\frac{D}{t} = r$$
$$r = \frac{D}{t}$$

33.
$$y = mx + \overset{\downarrow}{b}$$
$$y - mx = -mx + mx + b$$
$$y - mx = b \text{ or}$$
$$b = y - mx$$

35.
$$v = \frac{4\pi \overset{\downarrow}{r}^3}{3}$$
$$\frac{3}{4\pi} \cdot v = \frac{4\pi r^3}{3} \cdot \frac{3}{4\pi}$$
$$\frac{3v}{4\pi} = r^3 \text{ or}$$
$$r^3 = \frac{3v}{4\pi}$$

37.
$$\overset{\downarrow}{p}n = 1$$
$$\frac{pn}{n} = \frac{1}{n}$$
$$p = \frac{1}{n}$$

39.
$$A = \frac{1}{2} h \, (a + \overset{\downarrow}{b})$$
$$2A = \frac{2}{1} \cdot \frac{1}{2} h(a + b)$$
$$2A = h(a + b)$$
$$\frac{2A}{h} = \frac{h}{h} (a + b)$$
$$\frac{2A}{h} = a + b$$
$$\frac{2A}{h} - a = a - a + b$$
$$\frac{2A}{h} - a = b \text{ or}$$
$$b = \frac{2A}{h} - a$$

41.
$$a_n = \overset{\downarrow}{a_1} + (n - 1)d$$
$$a_n - (n - 1)d$$
$$= a_1 + (n - 1)d - (n - 1)d$$
$$a_n - (n - 1)d$$
$$= a_1 \text{ or}$$
$$a_1 = a_n - (n - 1)d$$

16

Section 1.4

43.

$$A = (\overset{\downarrow}{a} + b + c) / 3$$

$$3A = \frac{3}{1}(a + b + c) / 3$$

$$3A = a + b + c$$

$$3A - b = a + b - b + c$$

$$3A - b = a + c$$

$$3A - b - c = a + c - c$$

$$3A - b - c = a \text{ or}$$

$$a = 3A - b - c$$

45.

$$S = \frac{1}{2}n(\overset{\downarrow}{a_1} + a_n)$$

$$2S = \frac{2}{1} \cdot \frac{1}{2}n(a_1 + a_n)$$

$$2S = n(a_1 + a_n)$$

$$\frac{2S}{n} = \frac{n}{n}(a_1 + a_n)$$

$$\frac{2S}{n} = a_1 + a_n$$

$$\frac{2S}{n} - a_n = a_1 + a_n - a_n$$

$$\frac{2S}{n} - a_n = a_1 \text{ or}$$

$$a_1 = \frac{2S}{n} - a_n$$

47.

$$E = \frac{(\overset{\downarrow}{T_h} - T_c)}{T_h}$$

$$E \cdot T_h = T_h \cdot \frac{(T_h - T_c)}{T_h}$$

$$ET_h = T_h - T_c$$

$$ET_h + T_c = T_h - T_c + T_c$$

$$ET_h + T_c = T_h$$

$$ET_h - ET_h + T_c = T_h - ET_h$$

$$T_c = T_h - ET_h$$

49.

$$V = 344 + 0.6(\overset{\downarrow}{T} - 20)$$

$$V - 344 = 344 - 344 + 0.6(T - 20)$$

$$V - 344 = 0.6(T - 20) = \frac{6}{10}(T - 20)$$

$$\frac{10}{6}(v - 344) = \frac{10}{6} \cdot \frac{6}{10}(T - 20)$$

$$\frac{10}{6}(V - 344) = T - 20$$

$$\frac{10}{6}(V - 344) + 20 = T - 20 + 20$$

$$\frac{10}{6}(V - 344) + 20 = T \text{ or}$$

$$T = \frac{10}{6}(V - 344) + 20 \text{ or}$$

$$T = \frac{V - 344}{0.6} + 20$$

51.

$$D = rt; \quad r = \frac{D}{t}; \quad D$$

$$= \left(\frac{D}{t}\right)t; \quad D = D$$

53.

$$A = \frac{1}{2}h(a + b); \quad h = \frac{2A}{a + b}$$

$$A = \frac{1}{2}\left(\frac{2A}{a + b}\right)(a + b)$$

$$A = A$$

55.

$$y = 3x - 3; \quad x = \frac{1}{3}(y + 3)$$

$$y = 3\left[\frac{1}{3}(y + 3)\right] - 3$$

$$y = 3 \cdot \frac{1}{3}(y + 3) - 3$$

$$y = y + 3 - 3$$

$$y = 3$$

17

57. $y = 2x + 4; \; x = \frac{1}{2}(y - 4)$

$$y = 2\left[\frac{1}{2}(y - 4)\right] + 4$$

$$y = 2 \cdot \frac{1}{2}(y - 4) + 4$$

$$y = y - 4 + 4$$

$$y = y$$

59.

$$S = \frac{n}{2}\left(2a + (n - 1)\overset{\downarrow}{d}\right)$$

$$\frac{2}{n}S = \frac{2}{n} \cdot \frac{n}{2}(2a + (n - 1)d)$$

$$\frac{2S}{n} = 2a + (n - 1)d$$

$$\frac{2S}{n} - 2a = 2a - 2a + (n - 1)d$$

$$\frac{2S}{n} - 2a = (n - 1)d$$

$$\frac{\frac{2S}{n} - 2a}{(n - 1)} = \frac{(n - 1)d}{(n - 1)}$$

$$\frac{\frac{2S}{n} - 2a}{n - 1} = d \text{ or}$$

$$d = \frac{\frac{2S}{n} - 2a}{n - 1}$$

61.

$$y = -2 \text{ is below}$$

$$y = -2x + 3 \text{ for } x < 2.5;$$

$$-2 < -2x + 3$$

$$-2 + 2x < -2x + 2x + 3$$

$$-2 + 2x < 3$$

$$-2 + 2 + 2x < 3 + 2$$

$$2x < 5$$

$$\frac{2x}{2} < \frac{5}{2}$$

$$x < \frac{5}{2} \text{ or } x < 2.5$$

63.

$$y = 0 \text{ is below}$$

$$y = -2x + 3 \text{ for } x < 1.5$$

$$-2x + 3 > 0$$

$$-2x + 2x + 3 > 0 + 2x$$

$$3 > 2x$$

$$\frac{3}{2} > \frac{2x}{2}$$

$$\frac{3}{2} > x \Rightarrow x < \frac{3}{2} \text{ or}$$

$$x < 1.5$$

65.

$$y = x + 4 \text{ is above}$$

$$y = -\frac{1}{2}x \text{ for } x > -2$$

$$x + 4 > -\frac{1}{2}x + 1$$

$$x + \frac{1}{2}x + 4 > -\frac{1}{2}x + \frac{1}{2}x + 1$$

$$\frac{3}{2}x + 4 > 1$$

$$\frac{3}{2}x + 4 - 4 > 1 - 4$$

$$\frac{3}{2}x > -3$$

$$\frac{2}{3} \cdot \frac{3}{2}x > -3 \cdot \frac{2}{3}$$

$$x > -2$$

67.

$$y = 0 \text{ is below}$$

$$y = x + 4 \text{ for } x > -4$$

$$0 < x + 4$$

$$0 - 4 < x + 4 - 4$$

$$-4 < x \text{ or } x > -4$$

69.

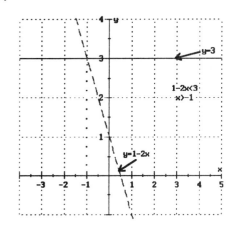

71.

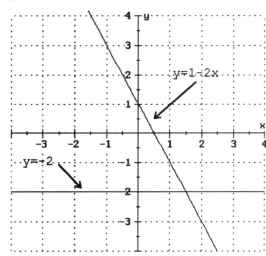

$$-2 < 1 - 2x, \quad x < \frac{3}{2}$$

73.

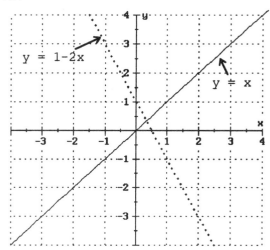

$$x < 1 - 2x, \quad x < \frac{1}{3}$$

75. $1 - 2x < x$

$1 - 2x + 2x < x + 2x$

$1 < 3x$

$\dfrac{1}{3} < \dfrac{3x}{3}$

$\dfrac{1}{3} < x \text{ or } x > \dfrac{1}{3}$

77.

$-\dfrac{1}{2}x + 1 < 4 - x$

$-\dfrac{1}{2}x + x + 1 < 4 - x + x$

$\dfrac{1}{2}x + 1 - 1 < 4 - 1$

$\dfrac{1}{2}x < 3$

$2 \cdot \dfrac{1}{2}x < 3 \cdot 2$

$x < 6$

Section 1.5

3. Cycles per second double each time. Not arithmetic.

5. a) $4 \,\diagdown\!\!\!/\, 8 \,\diagdown\!\!\!/\, 16 \,\diagdown\!\!\!/\, 20 \,\diagdown\!\!\!/\, 24$
           4   8   4   4

       not arithmetic

   b) $6 \,\diagdown\!\!\!/\, -1 \,\diagdown\!\!\!/\, -8 \,\diagdown\!\!\!/\, -15$
          -7   -7   -7

       arithmetic

   c) $2 \,\diagdown\!\!\!/\, 4 \,\diagdown\!\!\!/\, 8 \,\diagdown\!\!\!/\, 16 \,\diagdown\!\!\!/\, 32$
          2   4   8 16

       not arithmetic

   d)

   $\$20,000 \,\diagdown\!\!\!/\, \$22,000 \,\diagdown\!\!\!/\, \$24,200 \,\diagdown\!\!\!/\, \$26,620 \,\diagdown\!\!\!/\, \$29,282$
       $\$2,000$ $\$2,200$ $\$2,420$ $\$2,662$

not arithmetic

7. $4 \,\diagdown\!\!\!/\, 9 \,\diagdown\!\!\!/\, 14 \,\diagdown\!\!\!/\, 19 \,\diagdown\!\!\!/\, 24$
       5   5   5   5

   a) First term $a_1 = 4$, $d = 5$

   b) $a_{10} = 4 + (10 - 1) \cdot 5$
       $a_{10} = 4 + 9 \cdot 5$
       $a_{10} = 4 + 45 = 49$

   c) $a_n = 4 + (n - 1) \cdot 5$

   d) $y = 4 + (x - 1) \cdot 5$
       $y = 4 + 5x - 5$
       $y = 5x - 1$

   e) $a_{20} = 4 + (20 - 1) \cdot 5$
       $a_{20} = 4 + 19 \cdot 5 = 99$
       $S_{20} = \left(\dfrac{20}{2}\right)(4 + 99)$
       $S_{20} = 10 \cdot 103$
       $S_{20} = 1030$

9. $-1 \,\diagdown\!\!\!/\, 2 \,\diagdown\!\!\!/\, 5 \,\diagdown\!\!\!/\, 8 \,\diagdown\!\!\!/\, 11$
       3   3   3   3

   a) $a_1 = -1$; $d = 3$

   b) $a_{10} = -1 + (10 - 1) \cdot 3$
       $a_{10} = -1 + 9 \cdot 3$
       $a_{10} = 26$

   c) $a_n = -1 + (n - 1) \cdot 3$

9. d) $y = -1 + (x - 1) \cdot 3$
        $y = -1 + 3x - 3$
        $y = 3x - 4$

    e) $a_{20} = -1 + (20 - 1) \cdot 3$
        $a_{20} = -1 + 19 \cdot 3 = 56$
        $S_{20} = \left(\dfrac{20}{2}\right)(-1 + 56)$
        $S_{20} = 10 \cdot 55$
        $S_{20} = 550$

11. $2 \,\diagdown\!\!\!/\, 0 \,\diagdown\!\!\!/\, -2 \,\diagdown\!\!\!/\, -4 \,\diagdown\!\!\!/\, -6$
       -2 -2 -2 -2

   a) $a_1 = 2$; $d = -2$

   b) $a_{10} = 2 + (10 - 1) \cdot (-2)$
       $a_{10} = 2 + 9 \cdot (-2)$
       $a_{10} = -16$

   c) $a_n = 2 + (n - 1) \cdot (-2)$

   d) $y = 2 + (x - 1) \cdot (-2)$
       $y = 2 - 2x + 2$
       $y = -2x + 4$

   e) $a_{20} = 2 + (20 - 1) \cdot (-2)$
       $a_{20} = 2 + 19 \cdot (-2) = -36$
       $S_{20} = \left(\dfrac{20}{2}\right)(2 + (-36))$
       $S_{20} = 10 \cdot (-34)$
       $S_{20} = -340$

13. $1 \,\diagdown\!\!\!/\, -3 \,\diagdown\!\!\!/\, -7 \,\diagdown\!\!\!/\, -11 \,\diagdown\!\!\!/\, -15$
       -4   -4   -4   -4

   a) $a_1 = 1$; $d = -4$

   b) $a_{10} = 1 + (10 - 1) \cdot (-4)$
       $a_{10} = 1 + 9 \cdot (-4)$
       $a_{10} = -35$

   c) $a_n = 1 + (n - 1) \cdot (-4)$

   d) $y = 1 + (x - 1) \cdot (-4)$
       $y = 1 - 4x + 4$
       $y = -4x + 5$

13. e) $a_{20} = 1 + 19 \cdot (-4) = -75$

$$S_{20} = \left(\frac{20}{2}\right)(1 + (-75))$$

$$S_{20} = 10 \cdot (-74)$$

$$S_{20} = -740$$

19. None; y-intercept is f(0) and sequence starts at a, y-intercept is $a_1 - d$.

21.

x	$y = 2x + 3$	
1	2(1) + 3 = 5	
2	2(2) + 3 = 7	
3	2(3) + 3 = 9	
4	2(4) + 3 = 11	
5	2(5) + 3 = 13	

$1 <$ ... > 2 (repeated)

d = 2, arithmetic

23.

x	$y = 2x^2$	
1	$2(1)^2 = 2$	
2	$2(2)^2 = 8$	> 6
3	$2(3)^2 = 18$	> 10
4	$2(4)^2 = 32$	> 14
5	$2(5)^2 = 50$	> 18

Not arithmetic

25.

x	$y = 4 - 2x$	
1	4 − 2(1) = 2	
2	4 − 2(2) = 0	> −2
3	4 − 2(3) = −2	> −2
4	4 − 2(4) = −4	> −2
5	4 − 2(5) = −6	> −2

d = −2, arithmetic

27.

x	$y = 3 - x^2$	
1	$3 - (1)^2 = 2$	> −3
2	$3 - (2)^2 = -1$	> −5
3	$3 - (3)^2 = -6$	> −7
4	$3 - (4)^2 = -13$	> −9
5	$3 - (5)^2 = -22$	

Not arithmetic

29.

$a_1 = 1790; \; d = 10$

$a_n = 1790 + (n - 1)10$

$2000 = 1790 + (n - 1)10$

$2000 - 1790 = 1790 - 1790 + (n - 1)10$

$210 = (n - 1)10$

$\dfrac{210}{10} = (n - 1)\dfrac{10}{10}$

$21 = n - 1$

$21 + 1 = n - 1 + 1$

$22 = n$

22 times

31. $a_1 = 53; \; d = 1$

$a_n = 53 + (n-1)1$

$99 = 53 + (n-1)$

$99 = 52 + n$

$99 - 52 = 52 - 52 + n$

$47 = n$

47 tickets

33. 3 5 7 9 . . . ; $a_1 = 3$,

$d = 2$

$a_n = 3 + (n - 1)2 = 3 + 2n - 2$

$a_n = 2n + 1$

a) $a_1 = 1$

$a_n = 1 + (n-1)2 = 1 + 2n - 2$

$a_n = 2n - 1$

b) $a_1 = 5$

$a_n = 5 + (n-1)2 = 5 + 2n - 2$

$a_n = 2n + 3$

c) $a_1 = -1$

$a_n = -1 + 2n - 2$

$a_n = 2n - 3$

33. d) $a_1 = 2$

$a_n = 2 + 2n - 2$

$a_n = 2n$

CHAPTER ONE REVIEW

1. a) $f(0) = 1$
    b) $f(2) = 4$
    c) $f(-1) = \dfrac{1}{2}$
    d) $f(3) = 8$
    e) $f(1) = 2$
    f) $f(x) \geq 2$ for $x \geq 1$
    g) $f(x) < 0$ for no inputs
    h) $f(x) \leq 4$ for $x \leq 2$
    i) y-intercept = 1
    j) Domain $\mathbb{R}$
    k) Range $y > 0$

3. a) $\{1\}$
    b) $\{-4, 2\}$
    c) $\{-5, 3\}$
    d) $\{-3, 1\}$
    e) $\{ \ \}$; no solutions

5. a) $-8.9°\,C$
    b) $2.4°\ C$
    c) 4.0 km
    d) 1.0 km
    e) $15.7°\ C$

7 - 12.
   Answers for coordinate pair
   may vary. Answer shown is one
   possibility.

7. $(4, 16)$, $f(x) = x^2$

9. $(4, 8)$, $f(x) = 2x$

11. $(4, 4)$, $f(x) = 4$

13. $(0, 2)$ $(1, 3)$ $(2, 4)$
      $0 + 2 = 2$, $1 + 2 = 3$, etc.

15. $(0, 1)$ $(1, 2)$ $(2, 4)$
      $2^0 = 1$, $2^1 = 2$

17. $(0, 0)$ $(1, 3)$ $(2, 6)$
      $3(1) = 3$, $3(2) = 6$

19. $f(1) = 2(1)^2 - 3(1) + 1$
      $= 2 - 3 + 1 = 0$

21. $f(0.5) = 2(0.5)^2 - 3(0.5) + 1$
      $= 2(0.25) - 1.5 + 1$
      $= 0.5 - 1.5 + 1 = 0$

23. $f(-2) = 2(-2)^2 - 3(-2) + 1$
       $= 2(4) + 6 + 1 = 8 + 6 + 1 = 15$

25. $f(\square) = 2\square^2 - 3\square + 1$

27. $f(a) = 2a^2 - 3a + 1$

29. $f(a + b) = 2(a + b)^2 - 3(a + b) + 1$
       $= 2(a^2 + 2ab + b^2) - 3a - 3b + 1$
       $= 2a^2 + 4ab + 2b^2 - 3a - 3b + 1$

31. $f(x) = \sqrt{x + 2}$
       $x + 2 \geq 0$
       $x + 2 - 2 \geq 0 - 2$
       $x \geq -2$

33. $f(x) = \sqrt{2x - 1}$
       $2x - 1 \geq 0$
       $2x - 1 + 1 \geq 0 + 1$
       $2x \geq 1$
       $\dfrac{2x}{2} \geq \dfrac{1}{2}$
       $x \geq \dfrac{1}{2}$

35.

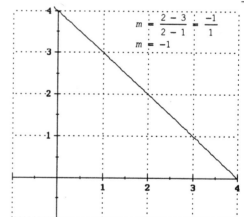

$m = \dfrac{2 - 3}{2 - 1} = \dfrac{-1}{1}$

$m = -1$

Chapter One Review

37.

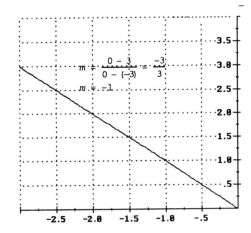

$$m = \frac{0 - 3}{0 - (-3)} = \frac{-3}{3}$$

$$m = -1$$

39.

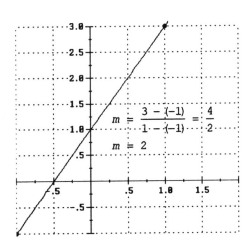

$$m = \frac{3 - (-1)}{1 - (-1)} = \frac{4}{2}$$

$$m = 2$$

41.

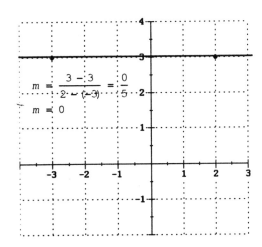

$$m = \frac{3 - 3}{2 - (-3)} = \frac{0}{5}$$

$$m = 0$$

43. a) # 41 is a horizontal line.

b) #35 & # 37 have equal slopes and are parallel lines.

c) #35: $y - 3 = -1(x - 1)$
$$y - 3 = -x + 1$$
$$y = -x + 1 + 3$$
$$y = -x + 4$$

#37: $y - 3 = -1(x - (-3))$
$$y - 3 = -x - 3$$
$$y = -x - 3 + 3$$
$$y = -x$$

#39: $y - (-1) = 2(x - (-1))$
$$y + 1 = 2x + 2$$
$$y = 2x + 2 - 1$$
$$y = 2x + 1$$

#41: $y - 3 = 0(x - (-3))$
$$y - 3 = 0$$
$$y = 3$$

45. a) $y = 0.065x$
b) m = 0.065; tax/$ purchased
y-intercept = 0;
0 tax on 0 purchase

47. a) y = $45x + $500
b) m=$45, Cost per hour of repairs
y-intercept = $500, inspection cost

49. a) y = $0.07x + $60
b) m = $0.07 cost per mile
y-intercept = $60
Cost if no miles are driven

51. a)

X	Y
1	4
2	6
3	8
4	10
5	12

b)

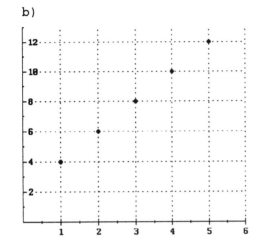

c) Do not connect,
   inputs are positive
   integers

d) $y = 2(x-1) + 4, \ x \geq 1$

53. $2\pi(1.5 \text{ in})^2 + 2\pi(1.5 \text{ in}) (4 \text{ in})$

$2\pi(2.25 \text{ in}^2) + 2\pi(6 \text{ in}^2)$

$2\pi(2.25 + 6) \text{ in}^2$

$2\pi(8.25) \text{ in}^2$

$16.5\pi \text{ in}^2 \text{ or} \approx 51.836 \text{ in}^2$

55. $20.00 + (8-1)(-\$1.75)$
     $20.00 + 7(-1.75)$
     $20.00 - 12.25$
     $\$7.75$

57. $2X + 2(1.5) = 11$

$2X + 3 = 11$

$2X + 3 - 3 = 11 - 3$

$2X = 8$

$\dfrac{2X}{2} = \dfrac{8}{2}$

$X = 4$

59. $40 = \dfrac{5}{9} (F - 32)$

$\dfrac{9}{5} \cdot 40 = \dfrac{9}{5} \cdot \dfrac{5}{9} (F - 32)$

$72 = F - 32$

$72 + 32 = F - 32 + 32$

$104 = F \text{ or}$

$F = 104$

61. $20 + (n - 1)(-1.75) = 0.75$

$20 - 20 + (n - 1)(-1.75) = 0.75 - 20$

$(n - 1)(-1.75) = -19.25$

$(n - 1)\dfrac{(-1.75)}{(-1.75)} = \dfrac{-19.25}{-1.75}$

$n - 1 = 11$

$n - 1 + 1 = 11 + 1$

$n = 12$

63. $I = P\overset{\downarrow}{r} t$

$\dfrac{I}{Pt} = \dfrac{Prt}{Pt}$

$\dfrac{I}{Pt} = r \text{ or}$

$r = \dfrac{I}{Pt}$

65. $c = a + \overset{\downarrow}{b} Y$

$c - a = a - a + bY$

$c - a = bY$

$\dfrac{c - a}{Y} = \dfrac{bY}{Y}$

$\dfrac{c - a}{Y} = b \text{ or}$

$b - \dfrac{c - a}{Y}$

67. $8 \diagdown_{/} 15 \diagdown_{/} 22 \diagdown_{/} 29 \diagdown_{/} 36$
         $7 \quad 7 \quad 7 \quad 7$

$36 + 7 = 43$

69. $32 \diagdown_{/} 16 \diagdown_{/} 8 \diagdown_{/} 4 \diagdown_{/} 2$
      $\div 2 \ \div 2 \ \div 2 \ \div 2$

$2 \div 2 = 1$

Chapter One Review

71. $a_1 = 4(1) - 3 = 4 - 3 = 1$
     $a_2 = 4(2) - 3 = 8 - 3 = 5$
     $a_3 = 4(3) - 3 = 12 - 3 = 9$
     $a_4 = 4(4) - 3 = 16 - 3 = 13$

73. $18 \diagdown_{-2}\diagup 16 \diagdown_{-2}\diagup 14 \diagdown_{-2}\diagup 12 \diagdown_{-2}\diagup 10$

     $a_1 = 18,\ d = -2$

     $a_n = 18 + (n - 1)(-2)$

     $a_n = 18 - 2n + 2$

     $a_n = -2n + 20$

75. $a_{20} = 18 + (20 - 1)(-2)$

     $a_{20} = 18 + 19(-2) = -20$

     $s_{20} = \dfrac{20}{2}(18 + (-20))$

     $s_{20} = 10(-2)$

     $s_{20} = -20$

77. $a_n = 22{,}000 + (n - 1)3{,}000$

     $109{,}000 = 22{,}000 + 3{,}000n - 3{,}000$

     $109{,}000 = 19{,}000 + 3{,}000n$

     $90{,}000 = 3{,}000n$

     $30 = n$

     30 oil changes

79. $a_n = a_1 + (n - 1)d$

     $a_n - a_1 = (n - 1)d$

     $\dfrac{a_n - a_1}{d} = n - 1$

     $\dfrac{a_n - a_1}{d} + 1 = n$ or

     $n = \dfrac{a_n - a_1}{d}$

81. a) Domain = positive natural
        numbers ≥ 1

     b) Range = $\{4,\ 6,\ 8,\ 10,\ \ldots\}$

83. a) Domain = $r > 0$
     b) Range = $A > 0$

Chapter One Test

1. a) $10\diagdown\diagup18\diagdown\diagup26\diagdown\diagup34\diagdown\diagup42,\quad 42 + 8 = 50$
   $\qquad 8\quad 8\quad 8\quad 8$

   arithmetic

   $a_{10} = 10 + (10 - 1)8 = 10 + 9 \cdot 8 = 82$

   $S_{10} = \dfrac{10}{2}(10 + 82)$

   $S_{10} = 5 \cdot 92 = 460$

   b) $4\diagdown\diagup10\diagdown\diagup18\diagdown\diagup28\diagdown\diagup40,\quad 40 + 14 = 54$
   $\qquad 6\quad 8\quad 10\quad 12$

   Not arithmetic

   c) $3\diagdown\diagup9\diagdown\diagup27\diagdown\diagup81,\quad 81 \cdot 3 = 243$
   $\qquad \cdot 3\quad \cdot 3\quad \cdot 3$

   Not arithmetic

   d) $-16\diagdown\diagup-9\diagdown\diagup-2\diagdown\diagup5\diagdown\diagup12,\quad 12 + 7 = 19$
   $\qquad 7\quad 7\quad 7\quad 7$

   Arithmetic

   $a_{10} = -16 + (10 - 1)7 = -16 + 9 \cdot 7 = 47$

   $S_{10} = \dfrac{10}{2}(-16 + 47)$

   $S_{10} = 5(13) = 155$

2. a) Not a function,
      input Ali has 2 outputs.

   b) Function

   c) Function

   d) Not a function,
      input 2 has 2 outputs.

3. a) $f(-2) = 3(-2)^2 - 2(-2) - 4$
      $\quad = 3 \cdot 4 + 4 - 4$
      $\quad = 12$

   b) $f(0) = 3(0)^2 - 2(0) - 4$
      $\quad = 0 - 0 - 4$
      $\quad = -4$

   c) $f(2) = 3(2)^2 - 2(2) - 4$
      $\quad = 3 \cdot 4 - 4 - 4$
      $\quad = 12 - 8$
      $\quad = 4$

4. a) $m = \dfrac{4 - 2}{-2 - 5} = \dfrac{2}{-7} = -\dfrac{2}{7}$

   b) slope is the same, $-\dfrac{2}{7}$

4. c) $y - 2 = -\dfrac{2}{7}(x - 5)$

   $\quad y - 2 = -\dfrac{2}{7}x + \dfrac{10}{7}$

   $\qquad y = -\dfrac{2}{7}x + \dfrac{10}{7} + \dfrac{14}{7}$

   $\qquad y = -\dfrac{2}{7}x + \dfrac{24}{7}$

5. a) Zero
   b) Negative, decreasing
   c) Linear
   d) Opposite
   e) Constant
   f) Positive integers.

6. a) $A = \dfrac{a + \overset{\downarrow}{b} + c}{3}$

   $3A = a + b + c$

   $b = 3A - a - c$

   b) $A = \dfrac{h}{2}(a + \overset{\downarrow}{b})$

   $\dfrac{2A}{h} = a + b$

   $\dfrac{2A}{h} - a = b$ or

   $b = \dfrac{2A}{h} - a$

7. a) $y = \$7.00x + \$2.50$

   b) $\$7.00$ per mile.

8. a) Input = minutes = x
      Output = cost = y

   b) $(1, \$0.13), \ (19, \$2.11)$

   c) $m = \dfrac{2.11 - 0.13}{19 - 1} = \dfrac{1.98}{18} = \$0.11$

   $y - 0.13 = 0.11(x - 1)$

   $y - 0.13 = .11x - 0.11$

   $y = 0.11x - 0.11 + 0.13$

   $y = 0.11x + 0.02$

9. a) $\{-4, 1\}$

   b) $\{-3, 0\}$

   c) $\{\ \}$, no solutions

10. $S_n = \dfrac{n}{2}\left(a_1 + a_n\right)$

$\dfrac{2S_n}{n} = a_1 + a_n$

$\dfrac{2S_n}{n} - a_1 = a_n \quad or \quad a_n = \dfrac{2S_n}{n} - a_1$

Section 2.0

1. $15^2 = 15 \cdot 15$
   $= 225$

3. $\sqrt{49} = \sqrt{7 \cdot 7} = 7$

5. $11^2 = 11 \cdot 11 = 121$

7. $\sqrt{169} = \sqrt{13 \cdot 13} = 13$

9. $\sqrt{196} = \sqrt{4 \cdot 49} = \sqrt{4} \cdot \sqrt{49}$
   $= 2 \cdot 7 = 14$

11. $\sqrt{75} \approx 8.660$

13. $\sqrt{40} \approx 6.325$

15. $\sqrt{200} \approx 14.142$

17. $3\sqrt{2} \approx 4.243$; #12

19. $10\sqrt{2} \approx 14.142$; #15

21. $3\sqrt{10} \approx 9.487$; #16

23. $\sqrt{150} = \sqrt{6 \cdot 25} = \sqrt{6} \cdot \sqrt{25} = 5\sqrt{6}$

25. $\sqrt{80} = \sqrt{16 \cdot 5} = \sqrt{16} \cdot \sqrt{5} = 4\sqrt{5}$

27. $\sqrt{32} = \sqrt{16 \cdot 2} = \sqrt{16} \cdot \sqrt{2} = 4\sqrt{2}$

29. $1 + \sqrt{2} \approx 2.414$

31. $\dfrac{1 - \sqrt{3}}{2} \approx -0.366$

33. $\dfrac{2 + \sqrt{8}}{2} = \dfrac{2 + \sqrt{4 \cdot 2}}{2} = \dfrac{2 + 2\sqrt{2}}{2}$
    $= 1 + \sqrt{2} \approx 2.414$

35. $\dfrac{3 - \sqrt{27}}{6} = \dfrac{3 - \sqrt{9 \cdot 3}}{6} = \dfrac{3 - 3\sqrt{3}}{6}$
    $= \dfrac{1 - \sqrt{3}}{2} \approx -0.366$

37. $\dfrac{4 + \sqrt{8}}{4} = \dfrac{4 + \sqrt{4 \cdot 2}}{4} = \dfrac{4 + 2\sqrt{2}}{4}$
    $= \dfrac{2 + \sqrt{2}}{2} = 1 + \dfrac{\sqrt{2}}{2} \approx 1.707$

39. $\dfrac{2 + \sqrt{2}}{2} = 1 + \dfrac{\sqrt{2}}{2} \approx 1.707$

41. $\dfrac{4\sqrt{8}}{4} = \sqrt{8} = \sqrt{4 \cdot 2} = 2\sqrt{2} \approx 2.828$

43. $x^2 = 225$
    $x = \pm\sqrt{225} = \pm\sqrt{9 \cdot 25}$
    $= \pm(3 \cdot 5) = \pm 15$

45. $x^2 - 121 = 0$; $x^2 = 121$
    $x = \pm\sqrt{121} = \pm 11$

47. $16x^2 - 48 = 0$
    $16x^2 = 48$
    $x^2 = 3$
    $x = \pm\sqrt{3}$

49. $5x^2 - 45 = 0$
    $5x^2 = 45$
    $x^2 = 9$
    $x = \pm\sqrt{9} = \pm 3$

51. $100 \text{ in}^2 = 4\pi r^2$
    $r^2 = \dfrac{100}{4\pi} = \dfrac{25}{\pi}$
    $r = \sqrt{\dfrac{25}{\pi}} = \dfrac{5}{\sqrt{\pi}} \approx 2.821 \text{ in}$

Section 2.0

53. a) $4^2 + 8^2 = x^2$

$16 + 64 = x^2$

$80 = x^2$

$\sqrt{80} = x$

$\sqrt{16 \cdot 5} = x$

$x = 4\sqrt{5} \approx 8.944$

b) $x^2 + 5^2 = 9^2$

$x^2 = 9^2 - 5^2$

$x^2 = 81 - 25$

$x^2 = 56$

$x = \sqrt{56}$

$x = \sqrt{4 \cdot 14}$

$x = 2\sqrt{14} \approx 7.483$

C) $x^2 = 8^2 + 6^2$

$x^2 = 64 + 36$

$x^2 = 100$

$x = \sqrt{100}$

$x = 10$

55. a) $x^2 = 8^2 + 8^2$

$x^2 = 2 \cdot 8^2$

$x^2 = \sqrt{2 \cdot 64}$

$x = 8 \cdot \sqrt{2} \approx 11.314$

b) $x^2 = 12^2 + 12^2$

$x^2 = 2 \cdot 12^2$

$x^2 = \sqrt{2 \cdot 12^2}$

$x = 12\sqrt{2} \approx 16.971$

55. c) $x^2 + x^2 = 8^2$

$2x^2 = 8^2$

$x^2 = \dfrac{8^2}{2}$

$x = \dfrac{8}{\sqrt{2}} = \left(\dfrac{\sqrt{2}}{\sqrt{2}}\right) \cdot \dfrac{8}{\sqrt{2}}$

$x = \dfrac{8\sqrt{2}}{2}$

$x = 4\sqrt{2} \approx 5.657$

57.

$h^2 + 4^2 = 8^2$

$h^2 = 8^2 - 4^2$

$h^2 = 64 - 16 = 48$

$h = \sqrt{48} = \sqrt{16 \cdot 3}$

$h = 4\sqrt{3} \text{ in.} \approx 6.928 \text{ in.}$

59.

$h^2 + \left(\dfrac{5}{2}\right)^2 = 5^2$

$h^2 = 5^2 - \dfrac{5^2}{2^2}$

$h^2 = \dfrac{4 \cdot 5^2}{4} - \dfrac{5^2}{4} = \dfrac{3 \cdot 5^2}{4}$

$h = \sqrt{\dfrac{3 \cdot 5^2}{4}} = \dfrac{5\sqrt{3}}{2} \approx 4.330\text{in}$

61.

$x^2 = 25^2 + 16^2$

$x^2 = 625 + 256 = 881$

$x = \sqrt{881}\text{ft} \approx 29.682\text{ft}$

29

63.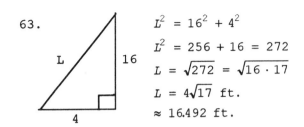

$L^2 = 16^2 + 4^2$

$L^2 = 256 + 16 = 272$

$L = \sqrt{272} = \sqrt{16 \cdot 17}$

$L = 4\sqrt{17}$ ft.

≈ 16.492 ft.

65. $z_1 = 5\sqrt{2}$, $z_2 = 10\sqrt{2}$;

$\dfrac{5\sqrt{2}}{5} = \sqrt{2} \approx 1.414, \dfrac{10\sqrt{2}}{10} = \sqrt{2} \approx 1.414$

Same for both triangles

67. $z^2 = x^2 + x^2$

$z^2 = 2x^2$

$z = x\sqrt{2}$;

$\dfrac{z}{x} = \dfrac{x\sqrt{2}}{x}$

$\dfrac{z}{x} = \sqrt{2}$

Section 2.1

1. $1 \cdot 48, 2 \cdot 24, 3 \cdot 16, 4 \cdot 12, 6 \cdot 8$

3. $1 \cdot 72, 2 \cdot 36, 3 \cdot 24, 4 \cdot 18, 6 \cdot 12, 8 \cdot 9$

5. $1 \cdot 90, 2 \cdot 45, 3 \cdot 30, 5 \cdot 18, 6 \cdot 15, 9 \cdot 10$

7.

m	n	$m+n$	$m \cdot n$
-3	-5	$-3 + (-5) = -8$	$-3 \cdot (-5) = 15$
3	4	$3 + 4 = 7$	$3 \cdot 4 = 12$
2	6	$2 + 6 = 8$	$2 \cdot 6 = 12$
3	5	$3 + 5 = 8$	$3 \cdot 5 = 15$
-4	-6	$-4 + (-6) = -10$	$-4 \cdot (-6) = 24$
-2	-12	$-2 + (-12) = -14$	$-2 \cdot (-12) = 24$
-3	-8	$-3 + (-8) = -11$	$-3 \cdot (-8) = 24$
2	-6	$2 + (-6) = -4$	$2 \cdot (-6) = -12$

9. a) $(x^3 + 2x^2 + 4x) - (2x^2 + 4x + 8)$
$$= x^3 + 2x^2 + 4x - 2x^2 - 4x - 8$$
$$= x^3 + 2x^2 - 2x^2 + 4x - 4x - 8$$
$$= x^3 - 8$$

b) $(x^3 - 6x^2 + 9x) - (3x^2 - 18x + 27)$
$$= x^3 - 6x^2 + 9x - 3x^2 + 18x - 27$$
$$= x^3 - 6x^2 - 3x^2 + 9x + 18x - 27$$
$$= x^3 - 9x^2 + 27x - 27$$

11. a) $x^2 + x^2 = 8^2$
$$2x^2 = 8^2$$
$$x^2 = \frac{8^2}{2}$$
$$(x^3 + 2x^2 + x) + (x^2 + 2x + 1)$$
$$= x^3 + 2x^2 + x + x^2 + 2x + 1$$
$$= x^3 + 2x^2 + x^2 + x + 2x + 1$$
$$= x^3 + 3x^2 + 3x + 1$$

b) $(a^3 - a^2b + ab^2) + (a^2b - ab^2 + b^3)$
$$= a^3 - a^2b + ab^2 + a^2b - ab^2 + b^3$$
$$= a^3 - a^2b + a^2b + ab^2 - ab^2 + b^3$$
$$= a^3 + b^3$$

13. a) $(x^3 + 2x^2 + 4x) = x(x^2 + 2x + 4)$; x

b) $2x^2 + 4x + 8 = 2(x^2 + 2x + 4)$; 2

15. a) $a^3 - a^2b + ab^2 = a(a^2 - ab + b^2)$; a

b) $a^2b - ab^2 + b^3$
$$= b(a^2 - ab + b^2) ; \; b$$

17. $x(x^2 - 2xy + y^2) - y(x^2 - 2xy + y^2)$
$$= x^3 - 2x^2y + xy^2 - x^2y + 2xy^2 - y^3$$
$$= x^3 - 2x^2y - x^2y + xy^2 + 2xy^2 - y^3$$
$$= x^3 - 3x^2y + 3xy^2 - y^3$$

19.

	$2x$	$+3$
$3x$	$6x^2$	$+9x$
-1	$-2x$	-3

$$= 6x^2 + 7x - 3$$

21.

	$3x$	$+1$
$3x$	$+9x^2$	$+3x$
$+1$	$+3x$	$+1$

$$= 9x^2 + 6x + 1$$

23.

	x	$+6$
x	x^2	$+6x$
-3	$-3x$	-18

$$= x^2 + 3x - 18$$

25.

	x	-6
x	x^2	$-6x$
$+3$	$+3x$	-18

$$= x^2 - 3x - 18$$

27.

	x	-9
x	x^2	$-9x$
-2	$-2x$	$+18$

$$= x^2 - 11x + 18$$

29. $(x + 4)(x - 4) = x^2 - 16$
(note: difference of squares)

31. $(x + 6)(x + 6) = x^2 + 2 \cdot 6x + 6^2$

$x^2 + 12x + 36$

33. $(x - 5)(x - 5) = x^2 + 2 \cdot (-5x) + (-5)^2$

$= X^2 - 10X + 25$

35.

	$2x$	-3
x	$2x^2$	$-3x$
$+4$	$+8x$	-12

$= 2x^2 + 5x - 12$

37.

	$2x$	$+3$
x	$2x^2$	$+3x$
-4	$-8x$	-12

$= 2x^2 - 5x - 12$

39.

	$3x$	-1
$3x$	$9x^2$	$-3x$
$+2$	$+6x$	-2

$= 9x^2 + 3x - 2$

41. $(x + 2)^2 = (x + 2)(x + 2)$

$= x^2 + 2 \cdot 2x + 4$

$= x^2 + 4x + 4$

43. $(2x - 1)(2x + 1) = 4x^2 - 1$

(note: difference of squares)

45.

factor	$6x$	$+1$
$2x$	$12x^2$	$+2x$
-3	$-18x$	-3

$= (2x - 3)(6x + 1)$

47.

factor	x	-3
$3x$	$3x^2$	$-9x$
-4	$-4x$	$+12$

$= (3x - 4)(x - 3)$

49. $x^2 + 7x + 12 = x^2 + 4x + 3x + 12$

	x	$+4$
x	x^2	$+4x$
$+3$	$+3x$	$+12$

$= (x + 3)(x + 4)$

51. $x^2 + x - 12 = x^2 + 4x - 3x - 12$

	x	$+4$
x	x^2	$+4x$
-3	$-3x$	-12

$= (x - 3)(x + 4)$

53. $x^2 - 7x + 12 = x^2 - 4x - 3x + 12$

	x	-4
x	x^2	$-4x$
-3	$-3x$	$+12$

$= (x - 3)(x - 4)$

55. $x^2 - 11x - 12 = x^2 - 12x + x - 12$

	x	-12
x	x^2	$-12x$
$+1$	x	-12

$= (x + 1)(x - 12)$

57. $x^2 + 3x - 28 = x^2 + 7x - 4x - 28$

	x	$+7$
x	x^2	$+7x$
-4	$-4x$	-28

$= (x - 4)(x + 7)$

59.

$6x^2 + 19x + 10 = 6x^2 + 4x + 15x + 10$

	$3x$	2
$2x$	$6x^2$	$+4x$
5	$15x$	$+10$

$= (3x + 2)(2x + 5)$

61. $6x^2 + 11x - 10 = 6x^2 + 15x - 4x - 10$

	$2x$	5
$3x$	$6x^2$	$+15x$
-2	$-4x$	-10

$= (2x + 5)(3x - 2)$

63. $6x^2 - 17x + 10 = 6x^2 - 12x - 5x + 10$

	x	-2
$6x$	$6x^2$	$-12x$
-5	$-5x$	$+10$

$= (6x - 5)(x - 2)$

65. $15x^2 - x - 6 = 15x^2 - 10x + 9x - 6$

	$3x$	-2
$5x$	$15x^2$	$-10x$
3	$9x$	-6

$= (5x + 3)(3x - 2)$

67. $6x^2 + 13x + 5 = 6x^2 + 3x + 10x + 5$

	$2x$	$+1$
$3x$	$6x^2$	$+3x$
$+5$	$+10x$	$+5$

$= (2x + 1)(3x + 5)$

69. $10x^2 + 61x + 6 = 10x^2 + 60x + x + 6$

	x	$+6$
$10x$	$10x^2$	$+60x$
$+1$	$+x$	$+6$

$= (10x + 1)(x + 6)$

71. $6x^2 - 32x + 10 = 2\left(3x^2 - 16x + 5\right)$

$= 2\left(3x^2 - 15x - x + 5\right)$

	x	-5
$3x$	$3x^2$	$-15x$
-1	$-x$	$+5$

$= 2(x - 5)(3x - 1)$

73. $y = \dfrac{-3}{40}\left(x^2 - 50x + 400\right)$

$-\dfrac{40}{3} y = x^2 - 50x + 400$

$-\dfrac{40}{3} y = x^2 - 10x - 40x + 400$

	x	-10
x	x^2	$-10x$
-40	$-40x$	$+400$

73. (cont)

$-\dfrac{40}{3} y = (x - 10)(x - 40)$

$y = -\dfrac{3}{40}(x - 10)(x - 40)$

x - intercepts are at

$x = 10$ & $x = 40$.

Vertex is halfway between these values at $x = 25$.

75. All are squares of binomials. Other like exercises #30, #31, #33.

77. Answers may vary

$12 = 12 \cdot 1,\ b = 12 + 1 = 13$

$12 = -12 \cdot (-1),\ b = -12 - 1 = -13$

$12 = 6 \cdot 2,\ b = 6 + 2 = 8$

$12 = -6 \cdot (-2),\ b = -6 - 2 = -8$

$12 = 3 \cdot 4,\ b = 3 + 4 = 7$

$12 = -3(-4),\ b = -3 - 4 = -7$

79. Answers may vary.

$-20 = -20 \cdot 1,\ b = -20 + 1 = -19$

$-20 = 20 \cdot (-1),\ b = 20 - 1 = 19$

$-20 = 2 \cdot (-10),\ b = 2 - 10 = -8$

$-20 = -2 \cdot (10),\ b = -2 + 10 = 8$

$-20 = 4 \cdot -5,\ b = 4 - 5 = -1$

$-20 = -4(5),\ b = -4 + 5 = 1$

81.

	x^2	$-2x$	$+4$
x	x^3	$-2x^2$	$+4x$
$+2$	$+2x^2$	$-4x$	$+8$

$x^3 + 8$

83. $(x - 2)(x^2 + 2x + 4)$

$= x^3 + 2x^2 + 4x - 2x^2 - 4x - 8$

$= x^3 - 8$

85. $12 = a(20 - 10)(20 - 30)$

$12 = a(10)(-10) = a(-100)$

$-\dfrac{12}{100} = a \ ; \quad \dfrac{-3 \cdot 4}{25 \cdot 4} = \dfrac{-3}{25} = a$

$y = \dfrac{-3}{25}(x - 10)(x - 30)$

Section 2.2

1. a) $7 \diagdown_{\diagup} 16 \diagdown_{\diagup} 27 \diagdown_{\diagup} 40 \diagdown_{\diagup} 55$
   $9 \diagdown_{\diagup} 11 \diagdown_{\diagup} 13 \diagdown_{\diagup} 15$
   $\quad 2 \quad 2 \quad 2$

   Quadratic

   b) $2a = 2;\ a = 1$
   $3a + b = 9;$
   $3(1) + b = 9,\ b = 6$
   $a + b + c = 7;$
   $1 + 6 + c = 7,\ c = 0$
   $f(x) = x^2 + 6x$

3. a) $4 \diagdown_{\diagup} 8 \diagdown_{\diagup} 12 \diagdown_{\diagup} 16 \diagdown_{\diagup} 20 \diagdown_{\diagup} 24$
   $\quad 4 \quad 4 \quad 4 \quad 4 \quad 4$

   Linear

   b) $a_n = 4 + (n - 1)4$

   $= 4 + 4n - 4 = 4n$

   $y = 4x$

5. a) $8 \diagdown_{\diagup} 27 \diagdown_{\diagup} 50 \diagdown_{\diagup} 77 \diagdown_{\diagup} 108$
   $19 \diagup 23 \diagdown 27 \diagdown 31$
   $\quad 4 \quad 4 \quad 4$

   Quadratic

   b) $2a = 4;\ a = 2$
   $3a + b = 19;\ 3(2) + b = 19,$
   $b = 13$
   $a + b + c = 8;$
   $2 + 13 + c = 8,\ c = -7$
   $f(x) = 2x^2 + 13x - 7$

7. a) $5 \diagdown_{\diagup} 12 \diagdown_{\diagup} 21 \diagdown_{\diagup} 32 \diagdown_{\diagup} 45$
   $7 \diagdown 9 \diagup 11 \diagup 13$
   $\quad 2 \quad 2 \quad 2$

   Quadratic

   b) $2a = 2;\ a = 1$
   $3a + b = 7;$
   $3(1) + b = 7,\ b = 4$
   $a + b + c = 5;$
   $1 + 4 + c = 5,\ c = 0$
   $f(x) = x^2 + 4x$

9. a) $13 \diagdown_{\diagup} 25 \diagdown_{\diagup} 37 \diagdown_{\diagup} 49 \diagdown_{\diagup} 61$
   $12 \quad 12 \quad 12 \quad 12$

   Linear

   b) $a_n = 13 + (n - 1)12$

   $= 13 + 12n - 12$

   $= 12n + 1,\ y = 12x + 1$

11. $1 \diagdown_{\diagup} 1 \diagdown_{\diagup} 2 \diagdown_{\diagup} 3 \diagdown_{\diagup} 5 \diagdown_{\diagup} 8$
    $0 \diagup 1 \diagup 1 \diagup 2 \diagup 3$
    $\quad 1 \quad 0 \quad 1 \quad 1$

    Neither

13. a) $9 \diagdown_{\diagup} 16 \diagdown_{\diagup} 21 \diagdown_{\diagup} 24 \diagdown_{\diagup} 25$
    $7 \diagdown 5 \diagup 3 \diagup 1$
    $\ -2 \quad -2 \quad -2$

    Quadratic

    b) $2a = -2;\ a = -1$
    $3a + b = 7;\ 3(-1) + b = 7,\ b = 10$
    $a + b + c = 9;\ -1 + 10 + c = 9,\ c = 0$
    $f(x) = -x^2 + 10x$

15. $A = \pi r^2 + 4\pi r$

    Input variable $= r$

    $a = \pi$

    $b = 4\pi$

    $c = 0$

17. $l = \dfrac{g}{4\pi^2}\, T^2$

    Input variable $= T$

    $a = \dfrac{g}{4\pi^2}$

    $b = 0$

    $c = 0$

19. $y = \dfrac{1}{2}\, x(x + 1) = \dfrac{1}{2}\, x^2 + \dfrac{1}{2}\, x$

    Input Variable $= x$

    $a = \dfrac{1}{2}$

    $b = \dfrac{1}{2}$

    $c = 0$

21. $y = (x - 1)^2 = x^2 - 2x + 1$

    Input Variable $= x$

    $a = 1$

    $b = -2$

    $c = 1$

Section 2.2

23. $y = 2(x + 1)^2 + 2$

    $= 2(x^2 + 2x + 1) + 2$

    $= 2x^2 + 4x + 2 + 2$

    $= 2x^2 + 4x + 4$

    Input Variable = x

    $a = 2 \quad b = 4 \quad c = 4$

25 – 27. Answers will vary. Possible
      answers are:

25. Bowling

27. Paint cans in a store window.

29. Calculator gives: $y = 0.032x^2$

    $$\frac{4}{125} = 0.032$$

31. Calculator gives:

    $y = -0.12x^2 + 4.8x - 36$

    $\dfrac{-12}{100} = -0.12, \ (x - 10)\,(x - 30)$

    $= x^2 - 40x + 300$

    $-0.12 \cdot (-40) = +4.8, \ -0.12 \cdot 300 = 36$

33. $h = -\dfrac{1}{2}gt^2 + v_0t + h_0$

    $v_0 = 2, \ h_0 = 35, \ g = 9.81, \ h = 0$

    $0 = -\dfrac{1}{2}(9.81)t^2 + 2t + 35$

    $0 = -4.905t^2 + 2t + 35$

35. $h_0 = 32.81, \ g = 32.2, \ v_0 = 4$

    $h = -\dfrac{1}{2}(32.2)t^2 + 4t + 32.81$

    $\quad = -16.1t^2 + 4t + 32.81$

    $v_0 = 8:$

    $h = -\dfrac{1}{2}(32.2)t^2 + 8t + 32.81$

    $\quad = -16.1t^2 + 8t + 32.81$

37. If $a = 0$ then $y = bx + c$ and
    this is a linear equation.

1. $2\sqrt{2} = \sqrt{4} \cdot \sqrt{2} = \sqrt{8}$

2. $5\sqrt{2} = \sqrt{25} \cdot \sqrt{2} = \sqrt{50}$

3. $4\sqrt{3} = \sqrt{16} \cdot \sqrt{3} = \sqrt{48}$

4. $6\sqrt{2} = \sqrt{36} \cdot \sqrt{2} = \sqrt{72}$

5. $3\sqrt{3} = \sqrt{9} \cdot \sqrt{3} = \sqrt{27}$

6. $2\sqrt{3} = \sqrt{4} \cdot \sqrt{3} = \sqrt{12}$

7. a) $-6 \diagdown\diagup -4 \diagdown\diagup 0 \diagdown\diagup 6 \diagdown\diagup 14 \diagdown\diagup \underline{\underline{24}}$
$\quad\ 2 \diagdown \diagup 4 \diagdown \diagup 6 \diagdown \diagup 8 \diagdown \diagup 10$
$\qquad\ 2 \quad 2 \quad 2 \quad 2$
Quadratic

$2a = 2;\ a = 1$
$3a + b = 2;\ 3(1) + b = 2,\ b = -1$
$a + b + c = -6$
$1 + (-1) + c = -6,\ c = -6$

$f(x) = x^2 - x - 6$

b) $3 \diagdown\diagup 0 \diagdown\diagup 1 \diagdown\diagup 6 \diagdown\diagup 15 \diagdown\diagup \underline{\underline{28}}$
$\ -3 \diagdown 1 \diagdown 5 \diagdown 9 \diagdown 13$
$\qquad 4 \quad 4 \quad 4 \quad 4$
Quadratic

$2a = 4;\ a = 2$
$3a + b = -3;\ 3(2) + b = -3,\ b = -9$
$a + b + c = 3$
$2 + (-9) + c = 3,\ c = 10$

$f(x) = 2x^2 - 9x + 10$

c) $8 \diagdown\diagup 11 \diagdown\diagup 14 \diagdown\diagup 17 \diagdown\diagup 20 \diagdown\diagup \underline{\underline{23}}$
$\quad\ 3 \quad 3 \quad 3 \quad 3 \quad 3$
Linear
$a_n = 8 + (n - 1)3 = 8 + 3n - 3$
$y = 3n + 5$

d) $3 \diagdown\diagup 4 \diagdown\diagup 7 \diagdown\diagup 11 \diagdown\diagup 18 \diagdown\diagup \underline{\underline{29}}$
$\quad\ 1 \diagdown 3 \diagdown 4 \diagdown 7 \diagdown 11$
$\qquad 2 \quad 1 \quad 3 \quad 4$
Other

8. a) $x^3 - 6x^2 + 9x - x^2 + 18x - 27$

$= x^3 - 6x^2 - x^2 + 9x + 18x - 27$

$= x^3 - 7x^2 + 27x - 27$

b) $x^3 + 3x^2 + 9x + (-3x^2 - 9x - 27)$

$= x^3 + 3x^2 + 9x - 3x^2 - 9x - 27$

$= x^3 + 3x^2 - 3x^2 + 9x - 9x - 27$

$= x^3 - 27$

c) $16a + 4b + c - (9a + 3b + c)$

$= 16a + 4b + c - 9a - 3b - c$

$= 16a - 9a + 4b - 3b + c - c$

$= 7a + b$

d) $9a + 3b + c - (4a + 2b + c)$

$= 9a + 3b + c - 4a - 2b - c$

$= 9a - 4a + 3b - 2b + c - c$

$= 5a + b$

9. a) $(x - 5)(x + 5) = x^2 - 25$

b) $(1 - x)(1 - x) = x^2 - 2x + 1$

c) $(1 - x)(x + 3) = x - x^2 + 3 - 3x$

$= -x^2 - 2x + 3$

d) $(2x + 3)(3 - 2x) = (3 + 2x)(3 - 2x)$

$9 - 4x^2 = -4x^2 + 9$

10. a) $3x^2 + 5x - 12 = 3x^2 + 9x - 4x - 12$

	x	$+3$
$3x$	$3x^2$	$+9x$
-4	$-4x$	-12

$= (3x - 4)(x + 3)$

b) $x^2 - x = x(x - 1)$

c) $6x^2 + 5x - 4 = 6x^2 + 8x - 3x - 4$

	$3x$	$+4$
$2x$	$6x^2$	$+8x$
-1	$-3x$	-4

$= (2x - 1)(3x + 4)$

d) $3x^3 + 6x^2 + 3x = 3x(x^2 + 2x + 1)$

$= 3x(x + 1)^2$

11. a) $x^2 + 5^2 = 13^2$

$\quad x^2 = 13^2 - 5^2 = 169 - 25$

$\quad x^2 = 144;\ x = 12$

b) $x^2 + 4^2 = 20^2$

$\quad x^2 = 20^2 - 4^2 = 400 - 16$

$\quad x^2 = 384$

$\quad x = \sqrt{384} = \sqrt{64 \cdot 6}$

$\quad x = 8\sqrt{6} \approx 19.596$

c) $x^2 = 2.5^2 + 6^2$

$\quad x^2 = 6.25 + 36 = 42.25$

$\quad x = \sqrt{42.25} = 6.5$

12. a) $7.5^2 + 10^2 \overset{?}{=} 12.5^2$

$\quad 56.25 + 100 \overset{?}{=} 156.25$

$\quad 156.25 = 156.25$

$\quad$ could be right triangle

b) $8^2 + 15^2 \overset{?}{=} 17^2$

$\quad 64 + 225 \overset{?}{=} 289$

$\quad 289 = 289$

$\quad$ could be right triangle

c) $6^2 + 8^2 \overset{?}{=} 12^2$

$\quad 36 + 64 \overset{?}{=} 144$

$\quad 100 \overset{?}{=} 144$

$\quad$ could not be right triangle

13.

Input	Output		
1	4		
2	12	> 8	> 4
3	24	> 12	> 4
4	40	> 16	> 4
5	60	> 20	

Quadratic;

13. (cont)

$\quad 2a = 4;\ a = 2$

$\quad 3a + b = 8:\ 3(2) + b = 8,\ b = 2$

$\quad a + b + c = 4:\ 2 + 2 + c = 4,\ c = 0$

$\quad f(x) = 2x^2 + 2x$

14. a)

x	$A = 6x^2$
0	$6(0)^2 = 0$
1	$6(1)^2 = 6$
2	$6(2)^2 = 6 \cdot 4 = 24$
3	$6(3)^2 = 6 \cdot 9 = 54$
4	$6(4)^2 = 6 \cdot 16 = 96$
5	$6(5)^2 = 6 \cdot 25 = 150$
6	$6(6)^2 = 6^3 = 216$
7	$6(7)^2 = 6 \cdot 49 = 294$
8	$6(8)^2 = 6 \cdot 64 = 384$
9	$6(9)^2 = 6 \cdot 81 = 486$
10	$6(10)^2 = 6 \cdot 100 = 600$

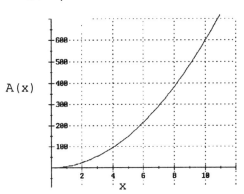

b) $x = 2,\ A = 24$

$\quad x = 4,\ A = 96;\ \dfrac{96}{24} = 4$

$\quad x = 8,\ A = 384;\ \dfrac{384}{96} = 4$

$\quad A = 6(2x)^2 = 6(4x^2) = 4(6x^2)$

$\quad$ 4 times original value.

Section 2.3

1. $1, 3, 9, 27, 81, 243, \underline{\underline{729}}$
   $2, 6, 18, 54, 162, 486$
   $4 \quad 12 \quad 36 \quad 108 \quad 324$
   Neither

3. $56, 47, 38, 29, 20, \underline{\underline{11}}$
   $-9 \quad -9 \quad -9 \quad -9 \quad -9$
   Linear

5. $2, 3, 6, 11, 18, \underline{\underline{27}}$
   $1, 3, 5, 7, 9$
   $\quad 2 \quad 2 \quad 2 \quad 2$
   Quadratic

7. x-intercepts (1, 0) & (3, 0)
   y-intercept (0, 3)
   vertex (2, -1)

9.

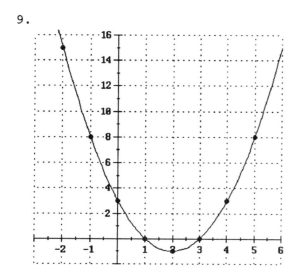

11. $x^2 - 4x + 3 = x^2 - 1x - 3x + 3$

	x	**-1**
x	x^2	$-x$
-3	$-3x$	$+3$

$= (x - 1)(x - 3)$

13. $f(6) = 15$ from symmetry of table.

15. $\{-1, 5\}$

17. $\{-2, 6\}$

19. $\{-5, 4\}$

21. $\{-2, 1\}$

23. a) (1, -9)

    b) (-2, 0) & (4, 0)

    c) (0, -8)

    d) {1}

    e) {-1, 3}

    f) {-2, 4}

    g) $y \geq -9$

25. a) (-1, 9)

    b) (-4, 0) and (2, 0)

    c) (0, 8)

    d) {-4, 2}

    e) {-2, 0}

    f) { }

    g) {-3, 1}

    h) $y \leq 9$

27.

r	$A = 2\pi r^2 + 12\pi r$
0	0
1	43.982
2	100.531
3	169.646
4	251.327
5	345.575
6	452.389

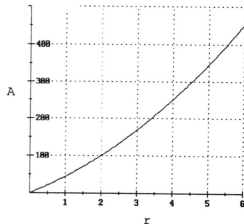

a) Answers will vary $r \approx 3.4$ in.

Section 2.3

27. b) Answers will vary $r \approx 5.5$ in.

c) Surface area varies with the square of the radius.

29. a) $6 - 2 = 4; \dfrac{4}{2} = 2$

$2 + 2 = 4$

$x = 4$

b) $6 - (-2) = 8; \dfrac{8}{2} = 4$

$-2 + 4 = 2$

$x = 2$

c) $4 - (-3) = 7; \dfrac{7}{2} = 3\dfrac{1}{2}$

$-3 + 3\dfrac{1}{2} = \dfrac{1}{2}$

$x = \dfrac{1}{2}$

31. $x = -4$

33. a) Vertex

b) x-intercept

c) y-intercept

35. a) x-intercept

b) Vertex

c) y-intercept

37. Initial or starting height.

39. ≈ -0.62 sec. has no meaning.

41. ≈ 1 sec.

43. ≈ 5.6 ft.

45. $0 = -16.1t^2 + 8t + 32.8$
$t \approx 1.697$ sec.
Which is ≈ 0.071 sec. longer
than original time of ≈ 1.626 sec.

47. a) $h = -\dfrac{1}{2}(32)t^2 + 60t + 0$

$h = -16t^2 + 60t$

b) $0 = -16t^2 + 60t = t(-16t + 60)$

$t = 0 \ or \ -16t + 60 = 0$

$-16t = -60$

$t = 3.75$

x-intercepts $(0, 0)$ and
$(3.75, 0)$

c)

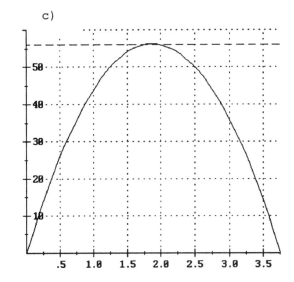

{1.75, 2} Reaches 56 ft. once on the way up and again on the way down.

d) Vertex at $x = \dfrac{3.75}{2} = 1.875$

$h = -16(1.875)^2 + 60(1.875)$

$h = 56.25$ft

Section 2.3

49. Answers will vary

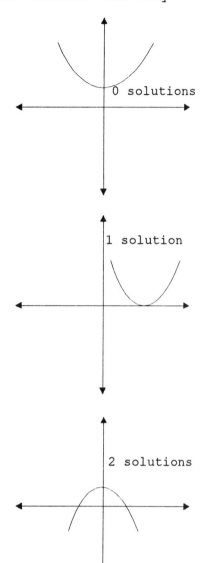

0 solutions

1 solution

2 solutions

51. $y = x - \dfrac{32.2}{2500} x^2$

This parabola opens downward and describes the path of a projectile.

53. a)

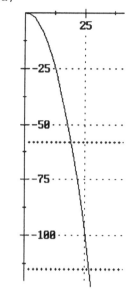

b) ≈ 19 ft.

c) ≈ 27 ft.

Section 2.4

1. $x^2 - x - 6 = 0$
    $(x - 3)(x + 2) = 0$
    $x - 3 = 0 \text{ or } x + 2 = 0$
    $x = 3 \text{ or } x = -2$
    $\{-2, 3\}$

3. $2x^2 + x - 1 = 0$
    $2x^2 + 2x - x - 1 = 0$
    $(2x - 1)(x + 1) = 0$
    $2x - 1 = 0 \text{ or } x + 1 = 0$
    $x = \dfrac{1}{2} \text{ or } x = -1$
    $\left\{-1, \dfrac{1}{2}\right\}$

5. $3x^2 - 48 = 0$
    $3x^2 = 48$
    $x^2 = 16$
    $x = \pm 4$
    $\{-4, 4\}$

7. $x^2 + 4x + 4 = 9$
    $x^2 + 4x - 5 = 0$
    $(x + 5)(x - 1) = 0$
    $x + 5 = 0 \text{ or } x - 1 = 0$
    $x = -5 \text{ or } x = 1$
    $\{-5, 1\}$

9. $x^2 - 10x + 25 = 36$
    $x^2 - 10x - 11 = 0$
    $(x + 1)(x - 11) = 0$
    $x + 1 = 0 \text{ or } x - 11 = 0$
    $x = -1 \text{ or } x = 11$
    $\{-1, 11\}$

11. $x^2 + 8x + 16 = 9$
    $x^2 + 8x + 7 = 0$
    $(x + 7)(x + 1) = 0$
    $x + 7 = 0 \text{ or } x + 1 = 0$
    $x = -7 \text{ or } x = -1$
    $\{-7, -1\}$

13. Solved using zoom & trace
    graphically: $\approx \{-0.45, 4.45\}$

15. $4x^2 - 1 = 0$
    $4x^2 = 1$
    $x^2 = \dfrac{1}{4}$
    $x = \pm \dfrac{1}{2}$
    $\left\{-\dfrac{1}{2}, \dfrac{1}{2}\right\}$

17. $4x^2 + 4x + 1 = 0$
    $4x^2 + 2x + 2x + 1 = 0$
    $(2x + 1)(2x + 1) = 0$
    $2x + 1 = 0$
    $2x = -1$
    $x = -\dfrac{1}{2}$
    $\left\{-\dfrac{1}{2}\right\}$

19. $3 = a(-4 + 1)(-4 + 5)$
    $3 = a(-3)(-1)$
    $3 = -3a$
    $-1 = a \text{ or } a = -1$

21. $7.5 = a(0 + 1)(0 + 5)$
    $7.5 = a(1)(5)$
    $5a = 7.5$
    $a = 1.5$

23. a)

$$0 = 75x^2 - 3 \quad \& \quad 0 = x^2 - \frac{3}{75}$$

$$3 = 75x^2 \qquad \frac{3}{75} = x^2$$

$$x = \pm\sqrt{\frac{3}{75}} = \pm\sqrt{\frac{1}{25}} = \pm\frac{1}{5}$$

x - intercepts for both graphs are

$$\left(-\frac{1}{5}, 0\right) \quad \& \quad \left(\frac{1}{5}, 0\right).$$

b)

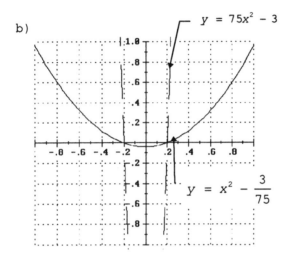

$$y = 75x^2 - 3$$

$$y = x^2 - \frac{3}{75}$$

c)   The intercepts are the solutions.

25.
$$y = a(x - 3)(x + 2)$$
$$12 = a(-3 - 3)(-3 + 2)$$
$$12 = a(-6)(-1)$$
$$12 = 6a$$
$$2 = a$$
$$f(x) = 2(x - 3)(x + 2)$$
$$f(x) = 2(x^2 + 2x - 3x - 6)$$
$$f(x) = 2x^2 - 2x - 12$$

27.
$$y = a(X - 3)(X + 2)$$
$$12 = a(0 - 3)(0 + 2)$$
$$12 = a(-3)(2)$$
$$12 = -6a$$
$$-2 = a$$
$$f(x) = -2(x - 3)(x + 2)$$
$$f(x) = -2(x^2 - x - 6)$$
$$f(x) = -2x^2 + 2x + 12$$

31. $l = (30 - 2w)/2$ line

33. Length and width
$\approx 13.5m$ and $\approx 1.5m$
Perimeter lines have the same slope.

35. $h = 0 \quad v_0 = 0 \quad 0 = \frac{-1}{2}gt^2 + h_0$

$$\frac{1}{2}gt^2 = h_0$$

$$t^2 = \frac{2h_0}{g}$$

$$t = \sqrt{\frac{2h_0}{g}}$$

37. $t = \sqrt{\frac{2(1368)}{32.2}} \cong 9.218 \text{ sec.}$

39. a) $A = \frac{\pi d^2}{4}, \quad \frac{4A}{\pi} = d^2$

$$d = \sqrt{\frac{4A}{\pi}} = 2\sqrt{\frac{A}{\pi}}$$

b) $\frac{\pi}{4} = 0.7854$

41. $d = 2\sqrt{\frac{5026 \text{ in}^2}{\pi}} \approx 80 \text{ in.}$

or $\approx 6\frac{2}{3}$ ft.

43. a) $r = \sqrt{12L}$

$$r^2 = 12L$$

$$L = \frac{r^2}{12}$$

Section 2.4

43. b) $L = \dfrac{55^2}{12} \approx 252\ ft.$

c) $r = \sqrt{12(100)} = \sqrt{1200}$

45. $y = \sqrt{(x - 2)^2}$
$r \approx 35\ \text{mph}$

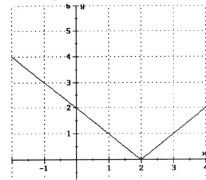

$y = |x - 2|$

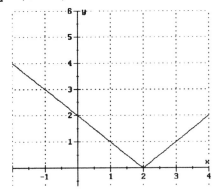

47. $y = \sqrt{(x - 1)^2}$

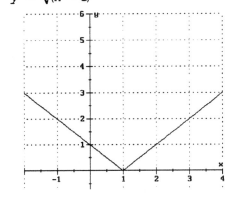

$y = |x - 1|$

49. $|x| = 4\ \{-4,\ 4\}$

51. $|x + 2| = 3;\ x + 2 = 3\ \text{or}\ x + 2 = -3,$
$x = 1\ \text{or}\ x = -5$
$\{-5, 1\}$

53. $|x - 5| = 2;\ x - 5 = 2\ \text{or}$
$x - 5 = -2,$
$x = 7\ \text{or}\ x = 3$
$\{3, 7\}$

55. $b\sqrt{a}$

57. $a\sqrt{b}$

59. $a^2\sqrt{b}$

1.

	a	$+7$
a	a^2	$7a$
$+7$	$7a$	$+49$

$$= a^2 + 14a + 49$$

3.

	x	± 8
x	x^2	$\pm 8x$
± 8	$\pm 8x$	$+64$

$$= x^2 + 16x + 64$$
$$\&\ x^2 - 16x + 64$$

5. $x^2 + 4x + \left(\dfrac{4}{2}\right)^2 = x^2 + 4x + 4$

7. $x^2 - 18x + \left(\dfrac{-18}{2}\right)^2 = x^2 - 18x + 81$

9. $x^2 - 7x + \left(\dfrac{-7}{2}\right)^2 = x^2 - 7x + \dfrac{49}{4}$

11. $x^2 + 12x + \left(\dfrac{12}{2}\right)^2 = 13 + \left(\dfrac{12}{2}\right)^2$

$x^2 + 12x + 36 = 49$

$(x + 6)^2 = 49$

13.

$x^2 - 11x + \left(\dfrac{-11}{2}\right)^2 = -18 + \left(\dfrac{-11}{2}\right)^2$

$x^2 - 11x + \dfrac{121}{4} = -18 + \dfrac{121}{4} = -\dfrac{72}{4} + \dfrac{121}{4}$

$\left(x - \dfrac{11}{2}\right)^2 = \dfrac{49}{4}$

15. $3x - 4x^2 = 5$

$0 = 4x^2 - 3x + 5$

$a_2 = 4,\ a_1 = -3,\ a_0 = 5$

17. $4 = 5 - x^2$

$x^2 - 1 = 0$

$a_2 = 1,\ a_1 = 0,\ a_0 = -1$

19. $5 - 3x^2 = 4$

$0 = 3x^2 - 1$

$a_2 = 3,\ a_1 = 0,\ a_0 = -1$

21.

$x^2 + 6x - 8 = 0 \quad a = 1,\ b = 6,\ c = -8$

$x = \dfrac{-b \pm \sqrt{b^2 - 4ac}}{2a},\ x = \dfrac{-6 \pm \sqrt{6^2 - 4 \cdot 1 \cdot (-8)}}{2 \cdot 1}$

$x = \dfrac{-6 \pm \sqrt{36 + 32}}{2} = \dfrac{-6 \pm \sqrt{68}}{2}$

$= \dfrac{-6 \pm \sqrt{4 \cdot 17}}{2} = \dfrac{-6 \pm 2\sqrt{17}}{2}$

$\left\{-3 - \sqrt{17},\ -3 + \sqrt{17}\right\};$

Real, 2 irrational number solutions.

23.

$x^2 + 6x + 8 = 0 \quad a = 1,\ b = 6,\ c = +8$

$x = \dfrac{-6 \pm \sqrt{6^2 - 4 \cdot 1 \cdot 8}}{2 \cdot 1}$

$x = \dfrac{-6 \pm \sqrt{36 - 32}}{2} = \dfrac{-6 \pm \sqrt{4}}{2}$

$= \dfrac{-6 \pm 2}{2} = -3 \pm 1$

$\{-4,\ -2\};$

Real, 2 rational number solutions

25.

$2x^2 + 3x = -1,\ 2x^2 + 3x + 1 = 0$

$a = 2,\ b = 3,\ c = +1$

$x = \dfrac{-3 \pm \sqrt{3^2 - 4 \cdot 2 \cdot 1}}{2 \cdot 2}$

$x = \dfrac{-3 \pm \sqrt{9 - 8}}{4} = \dfrac{-3 \pm 1}{4}$

$\left\{-1,\ -\dfrac{1}{2}\right\};$

Real, 2 rational number solutions

27. $2x^2 = -2 - 3x,\ 2x^2 + 3x + 2 = 0$

$a = 2,\ b = 3,\ c = +2$

$x = \dfrac{-3 \pm \sqrt{3^2 - 4 \cdot 2 \cdot 2}}{2 \cdot 2}$

$x = \dfrac{-3 \pm \sqrt{9 - 16}}{4} = \dfrac{-3 \pm \sqrt{-7}}{4}$

$\left\{\dfrac{-3 + \sqrt{-7}}{4},\ \dfrac{-3 - \sqrt{-7}}{4}\right\};$

No real number solution

Section 2.5

29.

$5x = 3 - 3x^2, \quad 3x^2 + 5x - 3 = 0$

$a = 3, \ b = 5, \ c = -3$

$x = \dfrac{-5 \pm \sqrt{5^2 - 4 \cdot 3 \cdot (-3)}}{2 \cdot 3}$

$x = \dfrac{-5 \pm \sqrt{25 + 36}}{6} = \dfrac{-5 \pm \sqrt{61}}{6}$

$\left\{ \dfrac{-5 - \sqrt{61}}{6}, \dfrac{-5 + \sqrt{61}}{6} \right\};$

Real, 2 irrational
number solutions

31.

$x = \dfrac{-(-4) - \sqrt{(-4)^2 - 4 \cdot (1) \cdot (-5)}}{2 \cdot (1)}$

$x = \dfrac{4 - \sqrt{16 + 20}}{2} = \dfrac{4 - \sqrt{36}}{2}$

$\dfrac{4 - 6}{2} = \dfrac{-2}{2} = -1$

$a = 1, \ b = -4, \ c = -5$

$x^2 - 4x - 5 = 0$

33. $x = \dfrac{-5 + \sqrt{5^2 - 4 \cdot (2) \cdot (-12)}}{2 \cdot (2)}$

$x = \dfrac{-5 + \sqrt{25 + 96}}{4} = \dfrac{-5 + \sqrt{121}}{4}$

$= \dfrac{-5 + 11}{4} = \dfrac{6}{4} = \dfrac{3}{2};$

$a = 2, \ b = 5, \ c = -12$

$2x^2 + 5x - 12 = 0$

35. $x = \dfrac{-1 + \sqrt{1^2 - 4 \cdot (3) \cdot (-4)}}{2 \cdot (3)}$

$x = \dfrac{-1 + \sqrt{1 + 48}}{6} = \dfrac{-1 + \sqrt{49}}{6}$

$\dfrac{-1 + 7}{6} = \dfrac{6}{6} = 1;$

$a = 3, \ b = 1, \ c = -4$

$3x^2 + x - 4 = 0$

37. $-0.12x^2 + 24x = 8$

$-0.12x^2 + 24x - 8 = 0$

$x = \dfrac{-24 \pm \sqrt{24^2 - 4 \cdot (-0.12) \cdot (-8)}}{2 \cdot (-0.12)}$

$x = \dfrac{-24 \pm \sqrt{5.76 - 3.84}}{-0.24} = \dfrac{-24 \pm \sqrt{1.92}}{-0.24}$

$\dfrac{-24 \pm \sqrt{.64 \cdot 3}}{-0.24} = \dfrac{-24 \pm .8\sqrt{3}}{-0.24}$

$x = \dfrac{-24}{-0.24} + \dfrac{.8\sqrt{3}}{-0.24}, \dfrac{-24}{-0.24} - \dfrac{.8\sqrt{3}}{-.24}$

$x = 10 - \dfrac{10\sqrt{3}}{3}, \ 10 + \dfrac{10\sqrt{3}}{3}$

$\left\{ \dfrac{30 - 10\sqrt{3}}{3}, \dfrac{30 + 10\sqrt{3}}{3} \right\}$ or

$\approx \{4.226, 15.774\}$

39. $-0.12x^2 + 24x = 10$

$-0.12x^2 + 24x - 10 = 0$

$x = \dfrac{-24 \pm \sqrt{24^2 - 4 \cdot (-0.12) \cdot (-10)}}{2 \cdot (-0.12)}$

$x = \dfrac{-24 \pm \sqrt{5.76 - 4.8}}{-0.24} = \dfrac{-24 \pm \sqrt{.96}}{-0.24}$

$\dfrac{-24 \pm \sqrt{0.16 \cdot 6}}{-0.24} = \dfrac{-24 \pm .4\sqrt{6}}{-0.24}$

$= \dfrac{30 \pm 5\sqrt{6}}{3}$

$\left\{ \dfrac{30 - 5\sqrt{6}}{3}, \dfrac{30 + 5\sqrt{6}}{3} \right\}$ or

$\approx \{5.918, 14.082\}$

41. $A = \pi r(4 + r) = 4\pi r + \pi r^2$

$A = \pi r^2 + 4\pi r; \text{ input} = r$

$a = \pi, \ b = 4\pi, \ c = 0$

43. $r = n(n + 1)$ $r = n^2 + n$

input = n,

$a = 1$, $b = 1$, $c = 0$

45. $A = P(1 + r)^2 = P(1 + 2r + r^2)$

$A = Pr^2 + 2Pr + P$, input = r

$a = p$, $b = 2P$, $c = P$

47. $h = -\dfrac{1}{2}(32.2)t^2 + 6t + 32.8$

$0 = -16.1t^2 + 6t + 32.8$

$t = \dfrac{-6 \pm \sqrt{6^2 - 4(-16.1)(32.8)}}{2(-16.1)}$

$= \dfrac{-6 \pm \sqrt{36 + 2112.32}}{-32.2} = \dfrac{-6 \pm \sqrt{2148.32}}{-32.2}$

$\dfrac{-6 + \sqrt{2148.32}}{-32.2} \approx -1.253$

(extraneous answer) or

$t = \dfrac{-6 - \sqrt{2148.32}}{-32.2} \approx 1.626$ sec.

49. $-16.1t^2 + 6t + 32.8 = 33$

$-16.1t^2 + 6t - .2 = 0$

$t = \dfrac{-6 \pm \sqrt{6^2 - 4(-16.1)(-.2)}}{2(-16.1)}$

$t = \dfrac{-6 \pm \sqrt{36 - 12.88}}{-32.2} = \dfrac{-6 \pm \sqrt{23.12}}{-32.2}$

$t \approx 0.037$ sec. or $t \approx 0.336$ sec.

51. $-16.1t^2 + 4t + 32.8 = 0$

$t = \dfrac{-4 \pm \sqrt{4^2 - 4(-16.1)(32.8)}}{2(-16.1)}$

$t = \dfrac{-4 \pm \sqrt{16 + 2112.32}}{-32.2} = \dfrac{-4 \pm \sqrt{128.32}}{-32.2}$

$t \approx -1.309$ (extraneous solution)

$t \approx 1.557$ sec.

53.

$-4.905t^2 + 2t + 10 = 5$

$-4.905t^2 + 2t + 5 = 0$

$t = \dfrac{-2 \pm \sqrt{2^2 - 4(-4.905)(5)}}{2(-4.905)}$

$t = \dfrac{-2 \pm \sqrt{102.1}}{-9.81}$

$t \approx 1.234$ sec.

($t \approx -0.826$ is an

extraneous answer)

55. $\sqrt{16x^2} = |4x|$

57. Cannot be simplified
Radicand does not contain
a perfect square factor.

59.

$2\square^2 + 3\square - 5 = 0$

$\square = \dfrac{-3 \pm \sqrt{3^2 - 4(2)(-5)}}{2(2)} = \dfrac{-3 \pm \sqrt{9 + 40}}{4}$

$= \dfrac{-3 \pm \sqrt{49}}{4} = \dfrac{-3 \pm 7}{4}$

$\square = \dfrac{-3 + 7}{4}$ or $\square = \dfrac{-3 - 7}{4}$

$\{\dfrac{-5}{2}, 1\}$

61.

$A^2 - 2A + 1 = 0$

$A = \dfrac{-(-2) \pm \sqrt{(-2)^2 - 4(1)(1)}}{2(1)}$

$= \dfrac{2 \pm \sqrt{4 - 4}}{2}$

$A = 1$; $A = x^2 = 1$

$x = \{-1, 1\}$

63. $x^2 = \dfrac{-(-8) \pm \sqrt{(-8)^2 - 4(16)(1)}}{2(16)}$

$= \dfrac{8 \pm \sqrt{64 - 64}}{32}$

$x^2 = \dfrac{8}{32} = \dfrac{1}{4}$; $x = \pm\sqrt{\dfrac{1}{4}}$

$x = \left\{-\dfrac{1}{2}, \dfrac{1}{2}\right\}$

CHAPTER TWO REVIEW

1. a) $\sqrt{75} = \sqrt{25 \cdot 3} = 5\sqrt{3}$

b) $\sqrt{8} = \sqrt{4 \cdot 2} = 2\sqrt{2}$

c) $\sqrt{32} = \sqrt{16 \cdot 2} = 4\sqrt{2}$

3. a) $\dfrac{3 - 3\sqrt{6}}{3} = 1 - \sqrt{6}$

b) $\dfrac{3 + 3\sqrt{6}}{3} = 1 + \sqrt{6}$

5. a) $-6 \quad -4 \quad 0 \quad 6 \quad 14 \quad 24$
   $\quad\quad 2 \quad 4 \quad 6 \quad 8 \quad 10$
   $\quad\quad\quad 2 \quad 2 \quad 2 \quad 2$

b) Quadratic

c) $2a = 2, \; a = 1$
   $3a + b = 2, \; 3(1) + b = 2, \; b = -1$
   $a + b + c = -6, \quad 1 - 1 + c = -6$
   $c = -6$
   $f(x) = x^2 - x - 6$

7. a) $-10 \quad -4 \quad 2 \quad 8 \quad 14 \quad 20$
   $\quad\quad 6 \quad 6 \quad 6 \quad 6 \quad 6$

b) Linear

c) $a_n = -10 + (n - 1)6 = -10 + 6n - 6$
   $f(x) = 6x - 16$

9. a) $(x - 3)(x + 3) = x^2 - 9$

b) $(2x - 5)(2x - 5) = 4x^2 - 20x + 25$

c) $(x - 1)(x^2 + x + 1)$

	x^2	x	1
x	x^3	x^2	x
-1	$-x^2$	$-x$	-1

$= x^3 - 1$

d) $(n + 4)(n + 4) = n^2 + 8n + 16$

11. a) $x^2 + 3x - 4 = x^2 + 4x - x - 4$
    $(x + 4)(x - 1)$

b) $2x^2 - 3x = x(2x - 3)$

c) $2x^2 + x - 3$
   $= 2x^2 + 3x - 2x - 3$

	2x	**+3**
x	$2x^2$	$+3x$
-1	$-2x$	-3

$(2x+3)(x-1)$

d) $9x^2 + 12x + 4 = (3x + 2)^2$

13.

r	$A = 4\pi r^2$
0	0
1	12.566
2	50.265
3	113.097
4	201.062
5	314.159
6	452.389
7	615.752
8	804.248
9	1017.876
10	1256.637

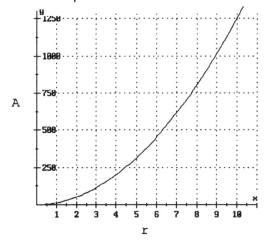

Chapter Two Review

15. a)

x	$f(x) = x^2 + x - 12$
-2	-10
-1	-12
0	-12
1	-10
2	-6

b) $0 = x^2 + x - 12$

$$x = \frac{-1 \pm \sqrt{1 - 4(1)(-12)}}{2(1)} = \frac{-1 \pm 7}{2}$$

$x = 3$ or $x = -4$

$(-4, 0)$ and $(3, 0)$

c) Vertex is between 0 and -1
at $x = -0.5$
$(-0.5)^2 + (-0.5) - 12 = -12.25$
$(-0.5, -12.25)$

d)

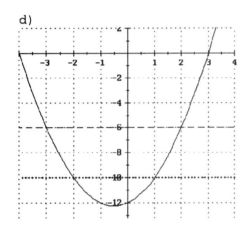

e) $x^2 + x - 12 = -6$

$\{-3, 2\}$

f) $x^2 + x - 12 = 0$

$\{-4, 3\}$

g) $x^2 + x - 12 = -10$

$\{-2, 1\}$

17. $T = 2\pi\sqrt{\dfrac{L}{g}}$

$$\frac{T}{2\pi} = \sqrt{\frac{L}{g}}$$

$$\frac{T^2}{4\pi^2} = \frac{L}{g}$$

$$L = \frac{gT^2}{4\pi^2}$$

19. True

21. True

23.

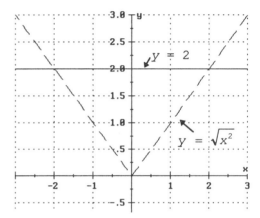

$x = \sqrt{4} \Rightarrow x = 2$

$\sqrt{x^2} = 2$

$x^2 = 4$

$x = \pm 2$

25.

$$x^2 + 3x - 4 = 0$$

$$x^2 + 4x - x - 4 = 0$$

$$(x + 4)(x - 1) = 0$$

$$(x + 4) = 0 \text{ or } (x - 1) = 0$$

$$x = -4 \quad x = 1 \quad \{-4, 1\}$$

$$x = \frac{-3 \pm \sqrt{3^2 - 4(1)(-4)}}{2(1)} = \frac{-3 \pm 5}{2}$$

$$x = \frac{-3 + 5}{2} = 1 \text{ or } x = \frac{-3 - 5}{2} = -4$$

$$\{-4, 1\}$$

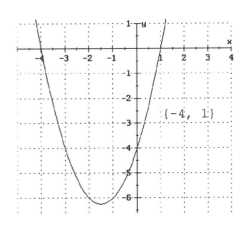

$\{-4, 1\}$

27. $3x^2 = 4 - x, \; 3x^2 + x - 4 = 0$

$3x^2 + 4x - 3x - 4 = 0$

	3x	+4
x	$3x^2$	+ 4x
-1	- 3x	-4

$(3x + 4)(x - 1) = 0$

$(3x + 4) = 0 \text{ or } (x - 1) = 0$

$3x = -4 \qquad\qquad x = 1$

$$x = \frac{-4}{3}; \quad \left\{-\frac{4}{3}, 1\right\}$$

$$x = \frac{-1 \pm \sqrt{1^2 - 4(3)(-4)}}{2(3)} = \frac{-1 \pm 7}{6}$$

$$x = \frac{-1 - 7}{6} = \frac{-8}{6} = \frac{-4}{3}$$

$$\text{or } x = \frac{-1 + 7}{6} = \frac{6}{6} = 1$$

$$\left\{\frac{-4}{3}, 1\right\}$$

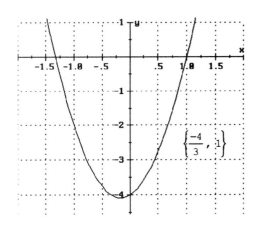

$\left\{\frac{-4}{3}, 1\right\}$

29. a) $l = 500\left|1 + \frac{8}{3}\left(\frac{30}{500}\right)^2\right|$

$l \approx 500|1 + 0.0096|$

$l \approx 500(1.0096) \approx 504.800ft$

b) $l = 500\left|1 + \frac{8}{3}\left(\frac{30}{500}\right)^2 - \frac{32}{5}\left(\frac{30}{500}\right)^4\right|$

$l \approx 500|1 + 0.0096 - 0.0000829|$

$l \approx 500(1.009517) \approx 504.7585ft$

c)

$l = 500\left|1 + \frac{8}{3}\left(\frac{30}{500}\right)^2 - \frac{32}{5}\left(\frac{30}{500}\right)^4 + \frac{256}{7}\left(\frac{30}{500}\right)^6\right|$

$l \approx 500|1 + 0.0096 - 0.0000829 + 0.00000171|$

$l \approx 500(1.009519) \approx 504.7594ft$

31. Second differences, quadratic regression

33.

$$\frac{x^2}{1.000 - x} = 4; \quad 0 \le x \le 1$$

$$x^2 = 4(1.000 - x) = 4.000 - 4x$$

$$x^2 + 4x - 4.000 = 0$$

$$x = \frac{-4 \pm \sqrt{4^2 - 4(1)(-4)}}{2(1)} = \frac{-4 \pm \sqrt{32}}{2}$$

$$= \frac{-4 \pm 4\sqrt{2}}{2} = -2 \pm 2\sqrt{2}$$

$$x = -2 - 2\sqrt{2} \approx -4.828$$

This is not a solution because of restriction on x, $0 \le x \le 1$

or $x = -2 + 2\sqrt{2} \approx \underline{0.828}$

35.

$$\frac{(3.000 + 2x)^2}{(1.000 - x)(2.000 - x)} = 42; \quad 0 \le x \le 1$$

$$\frac{9.000 + 12x + 4x^2}{2 - 3x + x^2} = 42$$

$$9 + 12x + 4x^2 = 42(2 - 3x + x^2)$$

$$9 + 12x + 4x^2 = 84 - 126x + 42x^2$$

$$0 = 38x^2 - 138x + 75$$

$$x = \frac{-(-138) \pm \sqrt{(-138)^2 - 4(38)(75)}}{2(38)}$$

$$x = \frac{138 \pm \sqrt{(-138)^2 - 11400}}{76}$$

$$x = \frac{138 \pm \sqrt{7644}}{76} = \frac{138 \pm \sqrt{196 \cdot 39}}{76}$$

$$= \frac{138 \pm 14\sqrt{39}}{76} = \frac{69 \pm 7\sqrt{39}}{38}$$

$$x = \frac{69 + 7\sqrt{39}}{38} \approx 2.966$$

Restriction on x eliminates this answer

$$\text{or } x = \frac{69 - 7\sqrt{39}}{38} \approx \underline{0.665}$$

37.

$$\iota \cdot \omega = 1.5 \text{ ft}^2 \quad 1.5 \text{ ft}^2 \left(\frac{144 \text{ in}^2}{1 \text{ ft}^2}\right) = 216 \text{ in}^2$$

$$\iota \cdot \omega = 216 \text{ in}^2$$

$$\frac{\iota}{\omega} = \frac{1.618}{1} \quad \iota = 1.618\omega$$

$$(1.618\omega)(\omega) = 1.618\omega^2 = 216$$

$$\omega^2 \approx 133.498 \quad \omega \approx 11.554 \text{ in.}$$

$$\iota = 1.618(11.554) \approx 18.695 \text{ in.}$$

39. Answers may vary
$$y = (x - 4)(x + 2)$$
$$y = x^2 - 2x - 8$$

41.
$$y = a(x - 4)(x + 2)$$
$$3 = a(1 - 4)(1 + 2)$$
$$3 = a(-3)(3)$$
$$3 = a(-9)$$
$$-\frac{1}{3} = a$$

$$y = -\frac{1}{3}(x - 4)(x + 2)$$

$$y = -\frac{1}{3}(x^2 - 2x - 8)$$

$$y = -\frac{1}{3}x^2 + \frac{2}{3}x + \frac{8}{3}$$

43.
a) Using linear regression on a calculator

$$y \approx 99.93 - 0.000989x$$

b) Using quadratic regression on a calculator

$$y \approx 0.00000000276x^2 - 0.00107x + 100.2$$

c) Both fit data very well

d) Linear $\Rightarrow y \approx 101.2°\,C$

Quadratic $\Rightarrow y \approx 101.6°\,C$

CHAPTER TWO TEST

1.

$$X = \frac{-b \pm \sqrt{b^2 - 4ac}}{2a}$$

2. a) $-6 \quad 2 \quad 12 \quad 24 \quad 38 \quad 54$
  $8 \quad 10 \quad 12 \quad 14 \quad 16$
  $2 \quad 2 \quad 2 \quad 2$

Quadratic

$2a = 2; \quad a = 1$
$3(1) + b = 8, \quad b = 5$
$(1) + (5) + c = -6, \quad c = -12$
$y = x^2 + 5x - 12$

b) $23 \quad 15 \quad 7 \quad -1 \quad -9 \quad -17$
  $-8 \quad -8 \quad -8 \quad -8 \quad -8$

Linear

$a_n = 23 + (n - 1)(-8)$
$a_n = 23 - 8n + 8$
$y = -8x + 31$

c) $2 \quad 8 \quad 14 \quad 20 \quad 26 \quad 32$
  $6 \quad 6 \quad 6 \quad 6 \quad 6$

Linear

$a_n = 2 + (n - 1)(6)$
$a_n = 2 + 6n - 6$
$y = 6x - 4$

d) $25 \quad 17 \quad 10 \quad 4 \quad -1 \quad -5$
  $-8 \quad -7 \quad -6 \quad -5 \quad -4$
  $+1 \quad +1 \quad +1 \quad +1$

Quadratic

$2a = 1; \quad a = \dfrac{1}{2}$

$3\left(\dfrac{1}{2}\right) + b = -8, \quad b = -\dfrac{19}{2}$

$\dfrac{1}{2} - \dfrac{19}{2} + c = 25, \quad c = 34$

$y = \dfrac{1}{2}x^2 - \dfrac{19}{2}x + 34$

3. $(3x - 4)(4x - 3)$

	3X	-4
4X	$12X^2$	-16X
-3	-9X	+12

$= 12X^2 - 25X + 12$

4. $(x - 3)(x^2 + 3x + 9)$

	x^2	+ 3x	+ 9
x	x^3	$3x^2$	9x
-3	$-3x^2$	- 9x	- 27

$= x^3 - 27$

5. $2x^2 - 7x + 6 = 2x^2 - 4x - 3x + 6$

	x	- 2
2x	$2x^2$	- 4x
- 3	- 3x	+ 6

$= (x - 2)(2x - 3)$

6. $0.04x^2 - 169 = 0$

$(0.2x - 13)(0.2x + 13) = 0$

$0.2x - 13 = 0 \ or \ 0.2x + 13 = 0$

$0.2x = 13 \qquad\qquad 0.2x = -13$

$x = 65 \qquad\qquad\quad x = -65$

$\{-65, 65\}$

7.

x	y
-8	18
-6	0
-4	-10
-2	-12
0	-6
2	8
4	30

a) $x^2 + 5x - 6 = -10, \quad x = -4$ from table
Other solution between 0 & -2
try -1

$(-1)^2 + 5(-1) - 6 \overset{?}{=} -10$
$1 - 5 - 6 = -10$
$\{-4, -1\}$

7. b) $x^2 + 5x - 6 = 8$, $x = 2$ from table

Other solution between -6 & -8

try -7

$(-7)^2 + 5(-7) - 6 \overset{?}{=} 8$

$49 - 35 - 6 = 8$

$\{-7, 2\}$

c) $x^2 + 5x - 6 = 0$, $x = -6$ from table

Other solution between 0 & 2

try 1

$(1)^2 + 5(1) - 6 \overset{?}{=} 0$

$1 + 5 - 6 = 0$

$\{-6, 1\}$

d) $x^2 + 5x - 6 = -20$

Lowest point between -1 & -4

try -3

$(-3)^2 + 5(-3) - 6 = 9 - 15 - 6 = -12$

try -2.5

$(-2.5)^2 + 5(-2.5) - 6 = -12.25$

Graph does not reach -20

no real number solution.

e) From (d) above, $(-2.5, -12.25)$

8. $2x^2 - 7x + 6 = 0$

$(x - 2)(2x - 3) = 0$

$x - 2 = 0$ or $2x - 3 = 0$

$x = 2 \qquad \qquad 2x = 3$

$\qquad \qquad \qquad x = \dfrac{3}{2}$

$\left\{ \dfrac{3}{2}, 2 \right\}$

9. $x = \dfrac{-(-7) \pm \sqrt{(-7)^2 - 4(2)(6)}}{2(2)}$

10. a)

r	$A = 20,000(1+r)^2$		
0.00	$20,000(1 + 0)^2$	=	20,000
0.01	$20,000(1 + 0.01)^2$	=	20,402
0.02	$20,000(1 + 0.02)^2$	=	20,808
0.03	$20,000(1 + 0.03)^2$	=	21,218
0.04	$20,000(1 + 0.04)^2$	=	21,632
0.05	$20,000(1 + 0.05)^2$	=	22,050
0.06	$20,000(1 + 0.06)^2$	=	22,472
0.07	$20,000(1 + 0.07)^2$	=	22,898
0.08	$20,000(1 + 0.08)^2$	=	23,328
0.09	$20,000(1 + 0.09)^2$	=	23,762
0.10	$20,000(1 + 010)^2$	=	24,200

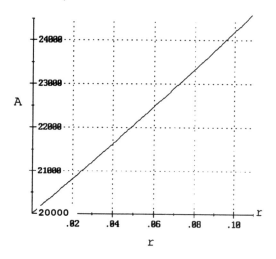

b) 7%

c) $\approx 22.5\%$

11. $v = \sqrt{2gs}$

$v^2 = 2gs$

$\dfrac{v^2}{2g} = s$

$s = \dfrac{v^2}{2g}$

12.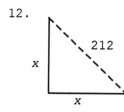

$x^2 + x^2 = 212^2$

$2x^2 = 212^2$

$x^2 \approx 22472$

$x \approx 150$ steps,

$\approx \underline{300\ steps}$

CHAPTER THREE

Section 3.0

1. $(x - 3)(x + 4) = x^2 + 4x - 3x - 12$
   $= x^2 + x - 12$

3. $(2x + 3)(x - 4) = 2x^2 - 8x + 3x - 12$
   $= 2x^2 - 5x - 12$

5. $(x + 3)(x + 3) = x^2 + 3x + 3x + 9$
   $= x^2 + 6x + 9$

7. $(x + 6)(x - 6) = x^2 - 6x + 6x - 36$
   $= x^2 - 36$

9. $(2x - 5)^2 = (2x - 5)(2x - 5)$
   $= 4x^2 - 10x - 10x + 25$
   $= 4x^2 - 20x + 25$

15. $x^2 + \left(2 \cdot \sqrt{16}\right)x + 16 = x^2 + \underline{8}x + 16$

17. $x^2 - \left(2 \cdot \sqrt{49}\right)x + 49 = \left(x - \sqrt{49}\right)^2$
    $= x^2 - \underline{14}x + 49 = \left(x - \underline{7}\right)^2$

19. $x^2 - 5x + (2.5)^2 = x^2 - 5x + \underline{6.25}$

21. $x^2 + 7x + \left(\dfrac{7}{2}\right)^2 = \left(x + \dfrac{7}{2}\right)^2$
    $x^2 + 7x + \underline{12.25} = (x + \underline{3.5})^2$

23. $x^2 - 24x + \left(\dfrac{24}{2}\right)^2 = \left(x - \dfrac{24}{2}\right)^2$
    $x^2 - 24x + \underline{144} = (x - \underline{12})^2$

25. $(x - 4)(x + 4) = x^2 + 4x - 4x - 16$
    $= x^2 - 16$

27. $(2x - 3)(2x + 3) = 4x^2 + 6x - 6x - 9$
    $= 4x^2 - 9$

31. $x^2 - 144 = \left(x + \sqrt{144}\right)\left(x - \sqrt{144}\right)$
    $(x + 12)(x - 12)$

33. $0.25x^2 - 0.01$
    $= \left(\sqrt{0.25}\,x + \sqrt{0.01}\right)\left(\sqrt{0.25}\,x - \sqrt{0.01}\right)$
    $= (0.5x + 0.01)(0.5x - 0.01)$

35. a) Set y = to 0, solve for x.

    b) $0 = (x + 1)^2$, $0 = x + 1$
       $-1 = x$, vertex at $(-1, 0)$

    c) $0 = (x - 4)^2$, $0 = x - 4$
       $4 = x$, vertex at $(4, 0)$

39. $(x + 1)(x^2 - x + 1)$

	x^2	$-x$	$+1$
x	x^3	$-x^2$	$+x$
$+1$	x^2	$-x$	$+1$

   $= x^3 + 1$
   or:
   $(x + 1)(x^2 - x + 1)$
   $= x^3 - x^2 + x + x^2 - x + 1$
   $= x^3 + 1$

41. $(x - 3)(x^2 - 6x + 9)$

	x^2	$-6x$	$+9$
x	x^3	$-6x^2$	$+9x$
-3	$-3x^2$	$+18x$	-27

   $= x^3 - 9x^2 + 27x - 27$
   or:
   $(x - 3)(x^2 - 6x + 9)$
   $= x^3 - 6x^2 + 9x - 3x^2 + 18x - 27$
   $= x^3 - 9x^2 + 27x - 27$

43. $(x + 1)(x^2 - 2x + 1)$

	x^2	$-2x$	$+1$
x	x^3	$-2x^2$	$+x$
+1	x^2	$-2x$	$+1$

$= x^3 - x^2 - x + 1$
or:
$(x + 1)(x^2 - 2x + 1)$
$= x^3 - 2x^2 + x + x^2 - 2x + 1$
$= x^3 - x^2 - x + 1$

47.

Factor	x^2	$+ 5x$	$+ 25$
x	x^3	$+5x^2$	$+25x$
−5	$-5x^2$	$-25x$	$- 125$

$x^3 - 125 = (x - 5)(x^2 + 5x + 25)$

49. $(x + 2)(x + 2)(x + 2)$

$= \big[(x + 2)(x + 2)\big](x + 2)$

$= \big(x^2 + 4x + 4\big)(x + 2)$

	x^2	$+4x$	$+4$
x	x^3	$+4x^2$	$+4x$
+2	$+2x^2$	$+8x$	$+8$

$= x^3 + 6x^2 + 12x + 8$
or:
$(x + 2)(x + 2)(x + 2)$

$= \big[(x + 2)(x + 2)\big](x + 2)$

$= \big(x^2 + 4x + 4\big)(x + 2)$

$= x^3 + 4x^2 + 4x + 2x^2 + 8x + 8$

$= x^3 + 6x^2 + 12x + 8$

Exercise 42

51. a) Perfect square trinomial

b) $(x^2 + 2x + 1) = (x + 1)^2$

c) $(x + 1)(x^2 + 2x + 1) = (x + 1)^3 = 0$
$x + 1 = 0, \ x = -1$

d) crosses once at $(-1, 0)$

57. $f(x) = x^2 - 4x \overset{\textcircled{\scriptsize 5}}{\longrightarrow} 5$ possible
solutions.

59. $f(x) = 2x \overset{\textcircled{\scriptsize 3}}{} - 2x^2 \longrightarrow 3$ possible
solutions.

61.

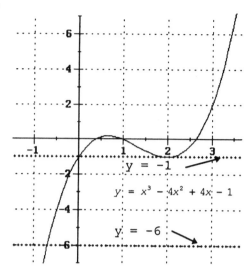

a) no solutions - not possible

b) One solution (answers may
vary) $y = -6$

c) Two solutions (answers may
vary) $y = -1$

d) Three solutions (answers may
vary) $y = 0$

e) Four solutions - not possible

Section 3.0

63.

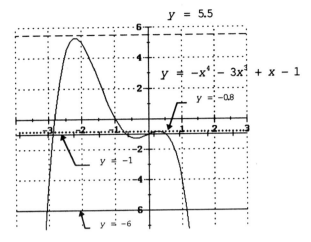

$y = 5.5$

$y = -x^4 - 3x^3 + x - 1$

$y = -0.8$

$y = -1$

$y = -6$

a) $y = 6$ (answers will vary)

b) $y \approx 5.5$

c) $y = -6$ (answers will vary)

d) $y \approx -0.8$ (answers will vary)

e) $y = -1$ (answers will vary)

Section 3.1

1. a is negative

3. $y = \dfrac{4}{125} x^2$; Positive coefficient on x^2 for parabola that opens upward.

5.

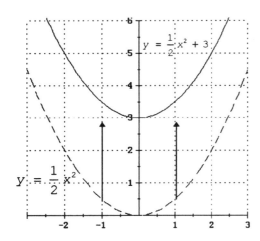

7.

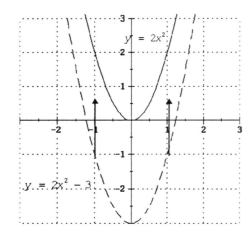

9-14 Put functions into standard form:

$$j(x) = 2(x - 1)^2 = 2(x^2 - 2x + 1)$$
$$= 2x^2 - 4x + 2$$

$$m(x) = \frac{x^2 + 2x}{2} = \frac{x^2}{2} + \frac{2x}{2}$$
$$= \frac{1}{2} x^2 + x$$

$$p(x) = -\frac{1}{2}(x + 1)^2 = -\frac{1}{2}(x^2 + 2x + 1)$$
$$= -\frac{1}{2} x^2 - x - \frac{1}{2}$$

$$q(x) = \frac{(x + 1)(x + 2)}{2} = \frac{x^2 + 3x + 2}{2}$$
$$\frac{1}{2} x^2 + \frac{3}{2} x + 1$$

9. $h(x)$, $m(x)$, $q(x)$

11. $f(x)$, $g(x)$, $h(x)$, $j(x)$, $m(x)$, $q(x)$

13. $j(x)$

15. $g(x)$

17. $j(x)$

19. Steeper

21. m

56

23.

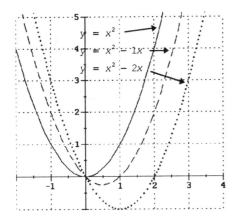

Vertex: $y = x^2$ $(0,\ 0)$

$$y = x^2 - 1x \quad \left(\frac{1}{2},\ -\frac{1}{4}\right)$$

$$y = x^2 - 2x \quad (1,\ -1)$$

25.

(200, 170)

(400, 20)

20 ft.

(0,0)

Fireboat

Fire

Nozzle (0,20)
Highpoint (200,170)
Fire (400, 20)
New equation has y-intercept =
20.

$$y = -0.00375x^2 + 1.5x + 20$$

Section 3.2

1. a) $\sqrt{-3} = \sqrt{-1}\sqrt{3} = i\sqrt{3}$

   b) $\sqrt{-54} = \sqrt{-1}\sqrt{54} = i\sqrt{9}\sqrt{6} = 3i\sqrt{6}$

   c) $\sqrt{-27} = \sqrt{-1}\sqrt{3^3} = 3i\sqrt{3}$

3. a) $16 = 16 + 0i$

   b) $\sqrt{-16} = \sqrt{-1}\sqrt{16} = 0 + 4i$

   c) $\sqrt{36} + \sqrt{-6} = 6 + \sqrt{-1}\sqrt{6}$
   $\qquad = 6 + i\sqrt{6}$

5. a) $\dfrac{6 + \sqrt{-12}}{2} = 3 + \dfrac{\sqrt{-1}\sqrt{12}}{2}$
   $\qquad = 3 + \dfrac{\sqrt{-1}\sqrt{4}\sqrt{3}}{2} = 3 + i\sqrt{3}$

   b) $\dfrac{2 - \sqrt{-24}}{4} = \dfrac{1}{2} - \dfrac{\sqrt{-1}\sqrt{24}}{4}$
   $\qquad = \dfrac{1}{2} - \dfrac{\sqrt{-1}\sqrt{4}\sqrt{6}}{4} = \dfrac{1}{2} - \dfrac{1}{2}i\sqrt{6}$

7. $2 - 3i$

9. $1 - i$

11. $-3 - 2i$

13. $2i + 3i - 4i - 5i + 6i + 7i - 8i - 9i$
    $= (2 + 3 - 4 - 5 + 6 + 7 - 8 - 9)i$
    $= 0 - 8i$

15. $6 + 4i - 7i + 5 - 12i$
    $= (6 + 5) + (4 - 7 - 12)i$
    $= 11 - 15i$

17. $-a + bi - (a + bi)$
    $= -a + bi - a - bi$
    $= -2a + 0i$

19. $3 - 4(5 - i) = 3 - 20 + 4i$
    $\qquad = -17 + 4i$

21. a) $(4 - 3i)(4 + 3i)$
    $= 16 - 12i + 12i - 9i^2$
    $= 16 - 9(-1) = 16 + 9 = 25$

    b) $(3 + 4i)(3 - 4i)$
    $= 9 + 12i - 12i - (16i^2)$
    $= 9 - 16(-1) = 9 + 16 = 25$

23. a) $(\sqrt{2} + i)(\sqrt{2} - i)$
    $= 2 + i\sqrt{2} - i\sqrt{2} - (i^2)$
    $2 - (-1) = 2 + 1 = 3$

    b) $(1 - i\sqrt{2})(1 + i\sqrt{2})$
    $= 1 - i\sqrt{2} + i\sqrt{2} - 2(i^2)$
    $1 - 2(-1) = 1 + 2 = 3$

25. $(x - 1)(x + 1) = x^2 - x + x - 1$
    $\qquad = x^2 - 1$

27. $(2 - \sqrt{2})(2 + \sqrt{2}) = 4 - 2\sqrt{2} + 2\sqrt{2} - 2$
    $\qquad = 4 - 2 = 2$

25 & 27: Results are the differences of two squares, with no middle terms or radicals. Changing $a - b$ to $b - a$ gives the negative of the original answer.

29. $(x + i)(x - i) = x^2 + xi - xi - i^2$
    $\qquad = x^2 - (-1) = x^2 + 1$

31. $(x - 3i)(x + 3i) = x^2 - 3xi + 3xi - 9i^2$
    $\qquad = x^2 - 9(-1) = x^2 + 9$

33. $2^2 - 4(1)(4) = 4 - 16 = -12$
    2 complex solutions;
    $$\dfrac{-2 \pm \sqrt{-12}}{2(1)} = \dfrac{-2 \pm 2i\sqrt{3}}{2}$$
    $\{-1 + i\sqrt{3}, -1 - i\sqrt{3}\}$

Section 3.2

35. $(-4)^2 - 4\,(1)\,(8) = 16 - 32 = -16$
   2 complex solutions;
   $$\frac{-(-4) \pm \sqrt{-16}}{2(1)} = \frac{4 \pm 4i}{2}$$
   $\{2 + 2i,\, 2 - 2i\}$

37. $x^3 + 8 = (x + 2)\,(x^2 - 2x + 4)$
   $x + 2 = 0$ or $x^2 - 2x + 4 = 0$
   $x = -2$ or
   $$x = \frac{-(-2 \pm \sqrt{(-2)^2 - 4(1)\,(4)}}{2(1)}$$
   $$= \frac{2 \pm \sqrt{-12}}{2} = \frac{2 \pm 2i\sqrt{3}}{2}$$
   $$= \left\{-2,\, 1 + i\sqrt{3},\, 1 - i\sqrt{3}\right\}$$

39. $x^4 - 81 = (x^2 - 9)\,(x^2 + 9)$
   $= (x^2 + 9)\,(x + 3)\,(x - 3)$
   $x^2 + 9 = 0$ or $x + 3 = 0$ or
   $x - 3 = 0$
   $x^2 = -9$ or $x = -3$ or
   $x = 3$
   $x = \pm\sqrt{-9}$
   $x = \pm 3i$
   $\{-3,\, 3,\, -3i,\, 3i\}$

41. $x^3 + 3x^2 + 2x = x(x^2 + 3x + 2)$
   $x = 0$ or $x^2 + 3x + 2 = 0$
   $\qquad\qquad (x + 2)\,(x + 1) = 0$
   $\qquad\qquad x + 2 = 0 \quad x = -2$
   $\qquad\qquad x + 1 = 0 \quad x = -1$
   $\{-2,\, -1,\, 0\}$

43. $\dfrac{x + 1}{2} = \dfrac{15}{x + 2}$
   $(x + 1)\,(x + 2) = 30$
   $x^2 + 3x + 2 = 30$
   $x^2 + 3x - 28 = 0$
   $$x = \frac{-3 \pm \sqrt{3^2 - 4 \cdot 1 \cdot (-28)}}{2 \cdot 1}$$
   $$x = \frac{-3 \pm \sqrt{9 + 112}}{2} = \frac{-3 \pm \sqrt{121}}{2}$$
   $$= \frac{-3 \pm 11}{2} \qquad x = \frac{-3 - 11}{2} = -7$$
   $$x = \frac{-3 + 11}{2} = 4$$
   $\{-7,\, 4\}$

45. $\dfrac{4}{x - 1} = \dfrac{x + 2}{10}$
   $40 = (x + 2)\,(x - 1)$;
   $40 = x^2 + x - 2$
   $x^2 + x - 42 = 0$
   $(x + 7)\,(x - 6) = 0$
   $x + 7 = 0 \quad x - 6 = 0$
   $x = -7 \qquad x = 6$
   $\{-7,\, 6\}$

47. $\dfrac{x + 1}{5} = \dfrac{4}{2x - 1}$
   $(x + 1)(2x - 1) = 20$
   $2x^2 + x - 1 = 20$
   $2x^2 + x - 21 = 0$
   $$x = \frac{-1 \pm \sqrt{(1)^2 - 4 \cdot 2 \cdot (-21)}}{2 \cdot 2}$$
   $$x = \frac{-1 \pm \sqrt{169}}{4} = \frac{-1 \pm 13}{4}$$
   $\{-3.5,\, 3\}$

49. True

51. True

53. c

55. b

57. a, e

Mid-Chapter 3 Test

1. a) $(2x + 3)(2x - 3)$

   $= 4x^2 + 6x - 6x - 9$

   $= 4x^2 - 9$

   Difference of squares

   b) $(2x - 3)(2x - 3)$

   $= 4x^2 - 6x - 6x + 9$

   $= 4x^2 - 12x + 9$

   Perfect square trinomial

   c) $(2x + 3)(2x + 3)$

   $= 4x^2 + 6x + 6x + 9$

   $= 4x^2 + 12x + 9$

   Perfect square trinomial

   d) $(3x - 2)(2x + 3)$

   $= 6x^2 - 4x + 9x - 6$

   $= 6x^2 + 5x - 6$

2. a) $(3 - 2i)(3 + 2i)$

   $= 9 - 6i + 6i - 4i^2$

   $= 9 - 4(-1) = 9 + 4 = 13$

   b) $(2 - i)(2 + i)$

   $= 4 - 2i + 2i - i^2$

   $= 4 - (-1) = 5$

   c) $(x - i)(x + i)$

   $= x^2 - xi + xi - i^2$

   $= x^2 - (-1) = x^2 + 1$

   d) $(2 - 3i)(2 + 3i)$

   $= 4 - 6i + 6i - 9i^2$

   $= 4 - 9(-1) = 13$

3. a) Results are both 13;
   Conjugates give real number
   products.

   b) $(1 + 2i)(1 - 2i)$

   $= 1 + 2i - 2i - 4i^2 = 5$

   $= (2 + i)(2 - i) = 5$ (from above)

4. $f(x) = -x^2 + 2x + 3 = 0$

   $2^2 - 4(-1)(3) = 4 + 12 = 16;$

   2 real solutions

5. $f(x) = (x - 2)(x^2 + 2x + 4) = 0$

   $x - 2 = 0$ or $x^2 + 2x + 4 = 0$

   $x = 2 \qquad$ or $x = \dfrac{-2 \pm \sqrt{(2)^2 - 4(1)(4)}}{2(1)}$

   $x = \dfrac{-2 \pm \sqrt{4 - 16}}{2}$

   $x = \dfrac{-2 \pm \sqrt{12}}{2}$

   $x = -1 \pm i\sqrt{3}$

   $\left\{2, -1 - i\sqrt{3}, \ -1 + i\sqrt{3}\right\}$

6.

	x^2	$2x$	$+ 4$
x	x^3	$2x^2$	$+ 4x$
$- 2$	$- 2x^2$	$- 4x$	$- 8$

   $= x^3 - 8$

7. Graph of $f(x) = (x - 2)(x^2 + 2x + 4)$
   has the same shape as $f(x) = x^3$
   but is moved down 8 units.

8. Graph moves down 4 units, vertex
   moves down 4 units, and
   x-intercepts at ± 2 instead of 0.

9. a) $x^2 + 4x + \left(\dfrac{4}{2}\right)^2 = x^2 + 4x + \underline{4}$

   b) $x^2 - 8x + \left(\dfrac{8}{2}\right)^2 = x^2 - 8x + \underline{16}$

   c) $x^2 + 3x + \left(\dfrac{3}{2}\right)^2 = x^2 + 3x + \dfrac{9}{4}$

10. Graph of $(x + 2)^2$ moves left 2
    units from $y = x^2$, vertex and
    x-intercept at $(-2, 0)$, and y-
    intercept at $(0, 4)$.

60

Section 3.3

1. $y = (x - 1)^2$ $h = 1, k = 0$
    vertex $= (1, 0)$

3. $y = (x + 3)^2 + 4$ $h = -3, k = 4$
    vertex $= (-3, 4)$

5. $y = (x - 2)^2 - 3$ $h = 2, k = -3$
    vertex $= (2, -3)$

7. Shift the graph of $y = x^2$
    2 units to the left.

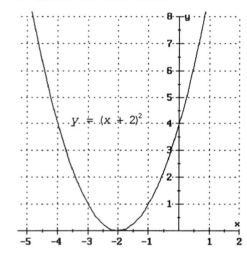

9. Shift the graph of $y = x^2$
    4 units to the right, and
    reflect over the x-axis.

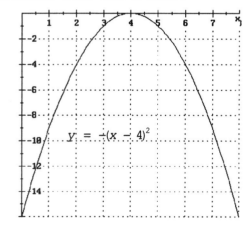

11. Shift the graph of $y = x^2$
     3 units to the right and
     4 units up.

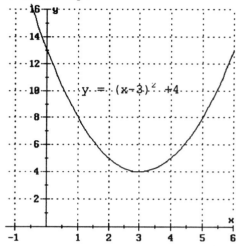

13. Shift the graph of $y = x^2$
     4 units to the left and 3 units
     up.

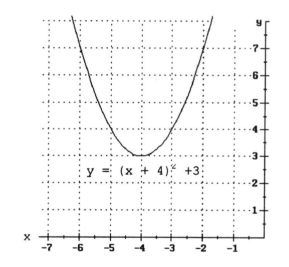

15. $y = x^2 + 10x + 25$ $\left(\dfrac{10}{2}\right)^2 = 25$

     $y = (x + 5)^2$

17. $y = x^2 - 6x + 9$ $\left(\dfrac{-6}{2}\right)^2 = 9$

     $y = (x - 3)^2$

61

19. $y = x^2 + 10x + 30$

    $y - 30 = x^2 + 10x$

    $\left(\dfrac{10}{2}\right)^2 = 25$

    $y - 30 + 25 = x^2 + 10x + 25$

    $y - 5 = (x + 5)^2$

    $y = (x + 5)^2 + 5$

21. $y = x^2 - 6x + 8$

    $y - 8 = x^2 - 6x \quad \left(\dfrac{-6}{2}\right)^2 = 9$

    $y - 8 + 9 = x^2 - 6x + 9$

    $y + 1 = (x - 3)^2$

    $y = (x - 3)^2 - 1$

23. a) This is $y = x^2$ shifted left 2 units. $h = -2$, $k = 0$. Equation is $y = (x + 2)^2$

    b) This is $y = x^2$ shifted right 3 units. $h = 3$, $k = 0$. Equation is $y = (x - 3)^2$

25. a) This is $y = x^2$ shifted left 1 unit and down 2 units. $h = -1$, $k = -2$. Equation is $y = (x + 1)^2 - 2$

    b) This is $y = x^2$ shifted right 2 units and up 1 unit. $h = 2$, $k = 1$. Equation is $y = (x - 2)^2 + 1$

27. a) One possible point is $(10,0)$
    $y = -0.24x^2 + 2.4x$

    b) $y = -0.24(x - 5)^2 + 6$

    $y = -0.24(x^2 - 10x + 25) + 6$

    $y = -0.24x^2 + 2.4x - 6 + 6$

    $y = -0.24x^2 + 2.4x$ ✓

29. a) Point $(0,0)$ becomes $(0, -25)$
    Point $(20, 10)$ becomes $(20, -15)$

    b) $y = 0.0375x^2 - 1.5x$

    or $y = 0.0375(x - 20)^2 - 15$

    c) Equations differ because the vertex of the new equation is at $(20, -15)$.

31. Start with $y = a(x - h)^2 + k$
    $h = 30$, $k = 100$ (vertex from problem statement)
    $y = a(x - 30)^2 + 100$
    origin is at $(0, 0)$
    $0 = a(0 - 30)^2 + 100$
    $-100 = a(0 - 30)^2$
    $-100 = a(-30)^2$
    $-100 = 900a$
    $-\dfrac{1}{9} = a;$
    $y = -\dfrac{1}{9}(x - 30)^2 + 100$

33. a) $y = |x + 2|$ is $y = |x|$ shifted left 2 units.

    b) Movement is the same.

    c) $y = (x + 2)^2$

35. a) $y = |x| - 2$ is $y = |x|$ shifted down 2 units.

    b) Movement is the same.

    c) $y = x^2 - 2$

Section 3.4

1. $y = x^2 - 2x = x(x - 2)$

   For intercepts $y = 0 = x(x - 2)$

   $x = 0$ or $x - 2 = 0$

   $x = 2$

   x - intercepts $(0,0)$, $(2,0)$.

   x coordinate of vertex is 1

   $y = 1(1 - 2) = -1$

   Vertex $= (1, -1)$.

3. $y = -x^2 - 4x + 21 = (-x - 7)(x - 3)$

   For intercepts $y = 0 = (-x - 7)(x - 3)$

   $-x - 7 = 0$ or $x - 3 = 0$

   $-x = 7$ $\qquad x = 3$

   $x = -7$

   x - intercepts $(-7, 0)$, $(3, 0)$.

   x - coordinate of vertex is halfway

   between -7 & 3. Distance between

   is 10 units so half way is 5 units;

   $-7 + 5 = -2$

   $y = (-(-2) - 7)(-2 - 3)$

   $y = (2 - 7)(-5)$

   $y = (-5)(-5) = 25$

   Vertex $= (-2, 25)$.

5. $f(x) = x^2 + 8x + 15$; $a = 1$, $b = 8$

   x - coordinate of vertex

   $= -\dfrac{8}{2 \cdot 1} = -4$

   $y = (-4)^2 + 8(-4) + 15 = -1$

   Vertex $(-4, -1)$

7. $f(x) = x^2 - 4x + 5$; $a = 1$, $b = -4$

   x coordinate of vertex

   $= -\dfrac{(-4)}{2 \cdot 1} = 2$

   $y = (2)^2 - 4(2) + 5 = 1$

   Vertex $(2, 1)$

9. $f(x) = 2x^2 - x - 3$; $a = 2$, $b = -1$

   x coordinate of vertex

   $= -\dfrac{(-1)}{2 \cdot 2} = \dfrac{1}{4}$

   $y = 2\left(\dfrac{1}{4}\right)^2 - \dfrac{1}{4} - 3 = \dfrac{2}{16} - \dfrac{4}{16} - 3$

   $\dfrac{1}{8} - \dfrac{2}{8} - \dfrac{24}{8} = -\dfrac{25}{8}$

   Vertex $\left(\dfrac{1}{4}, -\dfrac{25}{8}\right)$

11. Using $y = \dfrac{-g}{v^2} x^2 + x$;

    $y = -\dfrac{32.2}{90^2} x^2 + x$; $a = \dfrac{-32.2}{90^2}$, $b = 1$

    Max height is vertex y coordinate

    $x = -\dfrac{1}{2\left(-\dfrac{32.2}{90^2}\right)} = \dfrac{90^2}{2 \cdot 32.2}$

    $= \dfrac{90^2}{64.4} = 125.8$

    $y = \left(\dfrac{-32.2}{90^2}\right)\left(\dfrac{90^2}{2 \cdot 32.2}\right)^2 + \left(\dfrac{90^2}{2 \cdot 32.2}\right)$

    $y = -\dfrac{90^2}{4 \cdot 32.2} + \dfrac{90^2}{2 \cdot 32.2}$

    $= \dfrac{90^2}{4 \cdot 32.2} = 62.9$ ft.

    Maximum height ≈ 62.9 ft.

    Maximum distance to ground

    $y = 0 = x\left(\dfrac{-32.2}{90^2} x + 1\right)$

    $\dfrac{-32.2}{90^2} x + 1 = 0$ $\qquad \dfrac{-32.2}{90^2} x = -1$

    $x = -1\left(-\dfrac{90^2}{32.2}\right) \approx 251.6$ ft.

Section 3.4

13. $y = -\dfrac{32.2}{200^2}x^2 + x;\ a = \dfrac{-32.2}{200^2},\ b = 1$

$x = -\dfrac{1}{2\left(-\dfrac{32.2}{200^2}\right)} = \dfrac{200^2}{2 \cdot 32.2}$

$y = \left(\dfrac{-32.2}{200^2}\right)\left(\dfrac{200^2}{2 \cdot 32.2}\right)^2 + \left(\dfrac{200^2}{2 \cdot 32.2}\right)$

$y = -\dfrac{200^2}{4 \cdot 32.2} + \dfrac{200^2}{2 \cdot 32.2}$

$\quad = \dfrac{200^2}{4 \cdot 32.2} \approx 310.6\ \text{ft.}$

Does not reach 500 ft.

Maximum height $\approx$ 310.6 ft.

15. a)

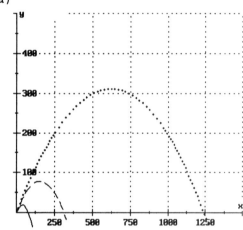

b) Quadrupled

c) Quadrupled

17. Answers will vary. One possible solution is a circle of circumference of 80 ft.
$80 = 2\pi r,\ r \approx 12.73\ \text{ft.}$

$A = \pi r^2 = \pi(12.73)^2 \approx 509\ \text{ft}^2$

19. $60 = \ell + 2w$

$\ell = 60 - 2w$

$A = \ell \cdot w = (60 - 2w)w$

$A = 60w - 2w^2 = -2w^2 + 60w$

Max area $\qquad$ from vertex

$w = -\dfrac{60}{2(-2)} = 15$

$60 = \ell + 2(15)$

$60 = \ell + 30$

$30 = \ell$

Length = 30 ft., Width = 15 ft.

21. Using $h = -0.5gt^2 + v_0 t + h_0$;

$v_0 = 40\ \text{ft/sec.},\quad g = 32.2\ \text{ft/sec}^2,$

$h_0 = 5.5\ \text{ft.}$

$h = -0.5(32.2)t^2 + 40t + 5.5$

$h = -16.1t^2 + 40t + 5.5$

$t \text{ at vertex} = \dfrac{-40}{2(-16.1)} = \dfrac{40}{32.2} \approx 1.24\ \text{sec.}$

$h \text{ at vertex} = -16.1(1.24)^2 + 40(1.24) + 5.5$

$h \approx 30.3\ \text{ft.}$

23.

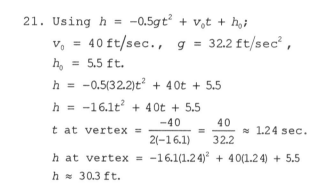

At $L\ x = 2000$

$Y = \dfrac{7}{400{,}000}(2000)^2 - \dfrac{1}{25}(2000) + 700 = 690$

Elevation at L = 690 ft.

Storm drain at vertex

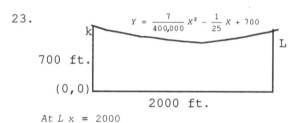

$x = -\dfrac{\left(\dfrac{-1}{25}\right)}{2 \cdot \dfrac{7}{400{,}000}} = \dfrac{1}{25} \cdot \dfrac{400{,}000}{14} \approx 1143$

$y = \dfrac{7}{400{,}000}(1143)^2 - \dfrac{1}{25}(1143) + 700 \approx 677$

Drain at $\approx (1143, 677)$

Section 3.4

25.

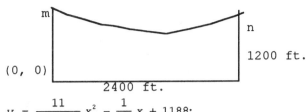

$$y = \frac{11}{480,000} x^2 - \frac{1}{20} x + 1188;$$

At m x = 0, m = 1188 ft.

Vertex at

$$x = -\frac{\left(\frac{-1}{20}\right)}{2 \cdot \frac{11}{480,000}} = \frac{1}{20} \cdot \frac{480,000}{22} \approx 1091$$

$$y = \frac{11}{480,000} (1091)^2 - \frac{1}{20} (1091) + 1188$$

$$y \approx 1161$$

Drain at (1091, 1161)

27. The other x-intercept is the same horizontal distance, $|h - x|$, from the vertex but on the opposite side.

Section 3.5

1. From graph and table in Ex. 4
   $2x^2 - 3x - 5 > 0$;
   when $x < -1$ or $x > 2.5$

3. From graph
   $\approx 12 < x < 100$
   (answer may vary slightly).

5. Highest point ≈ 28 *ft.*

7. 50 ft.

9. $x^2 - 5x + 6 = (x - 3)(x - 2)$
   Intercepts at $x = 3$, $x = 2$
   Graph turns up; ($a = 1$).

   $x^2 - 5x + 6 > 0$; $x < 2$ or $x > 3$

11. $x^2 - 5x - 14 = (x - 7)(x + 2)$
    Intercepts at $x = -2$, $x = 7$
    Graph turns up; ($a = 1$).

    $x^2 - 5x - 14 \leq 0$; $-2 \leq x \leq 7$

13. $x^2 - 5x - 6 = (x - 6)(x + 1)$
    Intercepts at $x = 6$, $x = -1$
    Graph turns up; ($a = 1$).

    $x^2 - 5x - 6 \geq 0$; $x \leq -1$ or $x \geq 6$

15. $x^2 + 6x + 9 = (x + 3)^2$
    Intercept at $x = -3$
    Graph turns up; ($a = 1$).

    $x^2 + 6x + 9 > 0$; $x < -3$ or $x > -3$

17. $4x^2 - 4x + 1$
    Intercept when $f(x) = 0$

    $x = \dfrac{-(-4) \pm \sqrt{(-4)^2 - 4 \cdot 4 \cdot 1}}{2(4)} = \dfrac{4 \pm \sqrt{16 - 16}}{8}$

    $x = \dfrac{1}{2}$
    Graph turns up; ($a = 4$).

    $4x^2 - 4x + 1 < 0$; $\{\ \}$ (never < 0).

19. c

21. f

23. n

25. x^2 is always ≥ 0,
    so if $y \geq x^2$, $y \geq 0$.

27. Answers will vary, some examples
    would be seasonal products such as
    ski equipment, heating oil, etc.

31. a) x-intercepts (120, 245)
       $b \approx 245 - 120 = 125$
       $h \approx 80,000$
       $A \approx \dfrac{2}{3}(125)(80,000) \approx 6,600,000$
       (Answer rounded down to nearest
       100,000)

    b) x-intercepts $\approx$ (0, 365)
       $b \approx 365$
       $h \approx 17,500$
       $A \approx \dfrac{2}{3}(365)(17,500) \approx 4,200,000$

    c) $6,600,000 - 4,200,000$
       $= 2,400,000 > 0$
       Yes, there is an overall profit.

33. x-intercepts (90, 120) and
    (270, 300)
    Vertex at 42,000 on both.

    b = 30, 30   h = 42,000
    Total Area $= \dfrac{2}{3}(30)(42,000)(2)$
    $A \approx 1,680,000$

    Total Production $\approx 1,680,000$ metric
    tons.

Section 3.5

35.   Sketch of Bridge

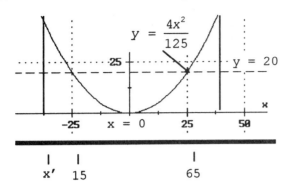

(Origin in center as equation has 0 as y-intercept)

$$20 = \frac{4x^2}{125} \quad 2500 = 4x^2$$

$$625 = x^2, \quad x = \pm 25$$

Team can inspect 25 feet on either side of the center. If you measure from the left end of the bridge: $15 < x < 65$, where $x = 0$ is at the left end of the bridge.

37. $\dfrac{1}{60{,}000} x^2 - \dfrac{3}{100} x + 1000 < 986.6;$

$\dfrac{1}{60{,}000} x^2 - \dfrac{3}{100} x + 13.4 < 0$

x - intercepts $\approx$

$$\frac{-\left(-\dfrac{3}{100}\right) \pm \sqrt{\left(-\dfrac{3}{100}\right)^2 - 4\left(\dfrac{1}{60{,}000}\right)(13.4)}}{2\left(\dfrac{1}{60{,}000}\right)}$$

$\approx \{822.5, \ 977.5\}$

$\dfrac{1}{60{,}000} x^2 - \dfrac{3}{100} x + 13.4 < 0;$

$822.5 < x < 977.5$

$977.5 - 822.5 = 155 \text{ ft.}$

1. $(x - 6)(x - 2)$

	x	$- 6$
x	x^2	$- 6x$
$- 2$	$- 2x$	$+ 12$

$= x^2 - 8x + 12;$

Other

3. $(x + 12)(x + 1)$

	x	$+ 12$
x	x^2	$+ 12x$
$+ 1$	$+ x$	$+ 12$

$= x^2 + 13x + 12;$

Other

5. $(x - 2)^2 = (x - 2)(x - 2)$

$= x^2 - 4x + 4;$

Perfect square trinomial

7. $x^2 - \underline{2\sqrt{225}}\, x + 225 = \left(x - \sqrt{225}\right)^2$

$x^2 - \underline{30x} + 225 = (x - \underline{15})^2$

9. $x^2 + 20x + \left(\dfrac{20}{2}\right)^2 = \left(x + \dfrac{20}{2}\right)^2$

$x^2 + 20x + \underline{100} = (x + \underline{10})^2$

11. $(a - b)(a^2 - 2ab + b^2)$

	a^2	$- 2ab$	$+ b^2$
a	a^3	$-2a^2b$	$+ ab^2$
$- b$	$- a^2b$	$+ 2ab^2$	$- b^3$

$= a^3 - 3a^2b + 3ab^2 - b^3$

13. $(a + b)(a^2 - ab + b^2)$

	a^2	$- ab$	$+ b^2$
a	a^3	$-a^2b$	$+ ab^2$
$+ b$	a^2b	$- ab^2$	$+ b^3$

$= a^3 + b^3$

15. $(a - b)^3 = (a - b)(a - b)(a - b)$

$= (a - b)(a^2 - 2ab + b^2)$

#11

17. Possible answers:

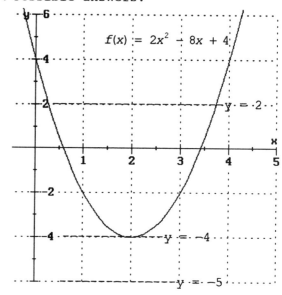

a) $y = -5$
b) $y = -4$
c) $y = 2$
d) & e) Not possible

19. $2x^2 - 8x + 4 = 0$

$x = \dfrac{-(-8) \pm \sqrt{(-8)^2 - 4(2)(4)}}{2(2)}$

$= \dfrac{8 \pm \sqrt{64 - 32}}{4}$

$= \dfrac{8 \pm \sqrt{32}}{4} = \dfrac{8 \pm \sqrt{16 \cdot 2}}{4}$

$\dfrac{8 \pm 4\sqrt{2}}{4} = 2 \pm \sqrt{2}$

$\left\{2 - \sqrt{2}, 2 + \sqrt{2}\right\} \approx \left\{0.586, \ 3.414\right\}$

21. Steeper

23. Shifted down 9 units

25. $y = x^2 - 9$ crosses x - axis twice,

$y = x^2 + 9$ not at all

27. Both factors are the same.

Chapter Three Review

29. $f(x) = -1x^2 + 2x - 1$

    $a = -1,\ b = 2,\ c = -1$

    $2^2 - 4(-1)(-1) = 4 - 4 = 0;$

    1 real number solution,

    1 x - intercept

31. $f(x) = 3x^2 + 2x + 1$

    $a = 3,\ b = 2,\ c = 1$

    $2^2 - 4 \cdot 3 \cdot 1 = 4 - 12 = -8;$

    No real number solutions,

    No x - intercepts.

33. c

35. b

37. b

39. a) $\sqrt{-16} = \sqrt{-1 \cdot 16} = 4i$

    b) $\sqrt{-50} = \sqrt{-1 \cdot 50} = i\sqrt{25 \cdot 2}$

       $= 5i\sqrt{2} \approx 7.071i$

41. $\dfrac{8 + 6i}{2} = \dfrac{8}{2} + \dfrac{6i}{2} = 4 + 3i$

43. a) $(4 + 3i)(4 + 3i)$

       $= 16 + 12i + 12i + 9i^2$

       $= 16 + 24i + 9 \cdot (-1)$

       $= 16 + 24i - 9 = 7 + 24i$

    b) $(4 - 3i)(4 + 3i)$

       $16 + 12i - 12i - 9i^2$

       $= 16 - 9i^2$

       $= 16 - 9(-1)$

       $= 16 + 9 = 25$

    c) $(3 - 4i)(3 + 4i)$

       $9 + 12i - 12i - 16i^2$

       $= 9 - 16i^2$

       $= 9 - 16(-1)$

       $= 9 + 16 = 25$

45. $f(x) = (x - 3)(x^2 + 3x + 9) = 0$

    $x - 3 = 0\ or\ x^2 + 3x + 9 = 0$

    $x = 3$

    $x = \dfrac{-3 \pm \sqrt{3^2 - 4 \cdot 1 \cdot 9}}{2 \cdot 1}$

    $= \dfrac{-3 \pm \sqrt{9 - 36}}{2}$

    $= \dfrac{-3 \pm \sqrt{-27}}{2}$

    $= \dfrac{-3 \pm 3i\sqrt{3}}{2}$

    $\left\{ 3,\ -\dfrac{3}{2} + \dfrac{3i\sqrt{3}}{2},\ -\dfrac{3}{2} - \dfrac{3i\sqrt{3}}{2} \right\}$

    $\approx \{3,\ -1.5 + 2.598i,\ -1.5\ -2.598i\}$

47. $f(x) = x^2 - x - 2$

    $= (x + 1)(x - 2)$

    x - intercepts $\{-1, 2\};$

    $-\dfrac{b}{2a} = -\left(\dfrac{-1}{2}\right) = \dfrac{1}{2} = x$

    $\left(\dfrac{1}{2}\right)^2 - \left(\dfrac{1}{2}\right) - 2 = \dfrac{1}{4} - \dfrac{1}{2} - 2$

    $= -2\dfrac{1}{4}$

    Vertex $\left(\dfrac{1}{2},\ -2\dfrac{1}{4}\right)$ or

    $(0.5,\ -2.25)$

49. $y = 2x^2 - 4x + 5$

    $a = 2,\ b = -4,\ c = 5$

    $(-4)^2 - 4 \cdot 2 \cdot 5 = 16 - 40 = -24$

    No x-intercepts;

    $-\dfrac{b}{2a} = -\dfrac{(-4)}{4} = 1$

    $2(1)^2 - 4(1) + 5$

    $= 2 - 4 + 5 = 3$

    Vertex $(1, 3)$

Chapter Three Review

51.

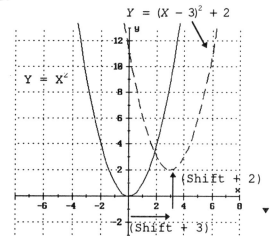

53.

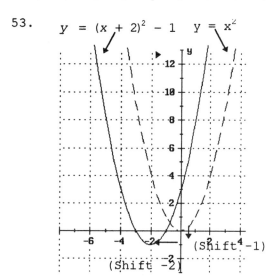

55. $y = x^2 + 4x + 9$

$y - 9 = x^2 + 4x \qquad \left[\left(\dfrac{4}{2} \right)^2 = 4 \right]$

$y - 9 + 4 = x^2 + 4x + 4$

$y - 5 = (x + 2)^2$

$y = (x + 2)^2 + 5$

57. a) $\{-2, 4\}$

b) $x = 1$

c) $(1, -27)$

59. Max height = vertex $(30, 100)$

$y = a(x - h)^2 + k$

$y = a(x - 30)^2 + 100$

(Jet starts at $(0,0)$)

$0 = a(0 - 30)^2 + 100$

$0 = a(-30)^2 + 100 = a(900) + 100$

$-100 = 900a \qquad a = -\dfrac{1}{9}$,

$y = -\dfrac{1}{9}(x - 30)^2 + 100$

$y = -\dfrac{1}{9}(x^2 - 60x + 900) + 100$

$y = -\dfrac{1}{9}x^2 + \dfrac{60}{9}x - 100 + 100$

$y = -\dfrac{1}{9}x^2 + \dfrac{60}{9}x = -\dfrac{1}{9}x^2 + \dfrac{20}{3}x$

$\approx -0.111x^2 + 6.667x$

61. $y = \dfrac{-g}{v^2}x^2 + x, \quad v = 132$

$y = \dfrac{-32.2}{(132)^2}x^2 + x$

$\dfrac{-b}{2a} = \dfrac{-1}{2\left(\dfrac{-32.2}{(132)^2}\right)} = \dfrac{(132)^2}{2(32.2)} \approx 271$

$y = \dfrac{-32.2}{(132)^2}\left(\dfrac{(132)^2}{2(32.2)}\right)^2 + \dfrac{(132)^2}{2(32.2)}$

$y = \dfrac{-132^2}{4 \cdot 32.2} + \dfrac{(132)^2}{2(32.2)}$

$= \dfrac{-132^2}{4 \cdot 32.2} + \dfrac{2(132)^2}{4(32.2)}$

$y = \dfrac{132^2}{4 \cdot 32.2} \approx 135$

Vertex $\approx (271, 135)$

63. $y = \dfrac{-32.2}{(400)^2} x^2 + x$

$\dfrac{-b}{2a} = \dfrac{-1}{2\left(\dfrac{-32.2}{(400)^2}\right)} = \dfrac{(400)^2}{2(32.2)} \approx 2484,$

$y = \dfrac{-32.2}{(400)^2}\left(\dfrac{(400)^2}{2(32.2)}\right)^2 + \dfrac{(400)^2}{2(32.2)}$

$y = \dfrac{-400^2}{4 \cdot 32.2} + \dfrac{(400)^2}{2(32.2)}$

$= \dfrac{-400^2}{4 \cdot 32.2} + \dfrac{2(400)^2}{4(32.2)}$

$y = \dfrac{400^2}{4 \cdot 32.2} \approx 1242$

Vertex $\approx (2484, 1242)$

65. $2\ell + 2w = 157$

$\ell = \dfrac{157 - 2w}{2}$

$A = \ell \cdot w = \left(\dfrac{157 - 2w}{2}\right)w$

$A = (78.5 - w)w$

$A = -w^2 + 78.5w;$

$\dfrac{-b}{2a} = -\dfrac{78.5}{2(-1)} = \dfrac{78.5}{2} = 39.25,$

$A = -(39.25)^2 + 78.5(39.25) \approx 1540.56$

Vertex $= (39.25, 1540.56)$

Max area for rectangular shape

is a square. $w = \ell \quad A = w^2$

$4w = 157 \quad w = 39.25$

$39.25^2 = 1540.56$

67. $x^2 + 6x - 8 \leq 0$ (graph turns up)

$x = \dfrac{-6 \pm \sqrt{(6)^2 - 4(1)(-8)}}{2(1)}$

$= \dfrac{-6 \pm \sqrt{36 + 32}}{2}$

$\dfrac{-6 \pm \sqrt{68}}{2} = \dfrac{-6 \pm 2\sqrt{17}}{2} = -3 \pm \sqrt{17}$

$-3 - \sqrt{17} \leq x \leq -3 + \sqrt{17}$

$\approx -7.123 \leq x \leq 1.123$

69. $1 - 4x^2 \leq 0$ (graph turns down)

$1 \leq 4x^2$

$\dfrac{1}{4} \leq x^2$

$0 \leq x^2 - \dfrac{1}{4}$

x - intercepts $\quad x^2 - \dfrac{1}{4} = 0$

$x^2 = \dfrac{1}{4}$

$x = \pm\dfrac{1}{2}$

$x \leq -\dfrac{1}{2}$ or $x \geq \dfrac{1}{2}$

71. $3x^2 + 5x - 3 > 19$ (graph turns up)

$3x^2 + 5x - 22 > 0$

$x = \dfrac{-5 \pm \sqrt{(5)^2 - 4(3)(-22)}}{2(3)}$

$= \dfrac{-5 \pm \sqrt{25 + 264}}{6}$

$\dfrac{-5 \pm \sqrt{289}}{6} = \dfrac{-5 \pm 17}{6}$

$x < \dfrac{-5 - 17}{6}$ or $x > \dfrac{-5 + 17}{6}$

$x < \dfrac{-22}{6}$ or $x > \dfrac{12}{6}$

$x < \dfrac{-11}{3}$ or $x > 2$

Chapter Three Review

73. $0.0375(x - 20)^2 + 10 < 20$

$0.0375(x^2 - 40x + 400) - 10 < 0$

$0.0375x^2 - 1.5x + 15 - 10 < 0$

$0.0375x^2 - 1.5x + 5 < 0$

$x^2 - 40x + \dfrac{400}{3} < 0$

$x = \dfrac{-(-40) \pm \sqrt{(-40)^2 - 4 \cdot 1 \cdot \dfrac{400}{3}}}{2 \cdot 1}$

$= \dfrac{40 \pm \sqrt{1600 - \dfrac{1600}{3}}}{2} = \dfrac{40 \pm \sqrt{\dfrac{3200}{3}}}{2}$

$= \dfrac{40 \pm 32.66}{2} \approx 20 \pm 16.33$

$3.67 < x < 36.33$ ft.

75. $\qquad y = a(x - 30)^2 + 100$

$0 = a(0 - 30)^2 + 100$

$0 = (900)a + 100$

$-100 = 900a \quad a = -\dfrac{1}{9},$

$y = -\dfrac{1}{9}(x - 30)^2 + 100$

$y = -\dfrac{1}{9}x^2 + \dfrac{60}{9}x$

$y = -\dfrac{1}{9}x^2 + \dfrac{60}{9}x < 50$

$y = -\dfrac{1}{9}x^2 + \dfrac{60}{9}x - 50 < 0$

$x^2 - 60x + 450 > 0$

$x = \dfrac{+60 \pm \sqrt{(-60)^2 - 4(1)\,(450)}}{2(1)}$

$x = \dfrac{60 \pm \sqrt{1800}}{2} = \dfrac{60 \pm 30\sqrt{2}}{2}$

$\dfrac{60 + 30\sqrt{2}}{2} \approx 51.213$

$\dfrac{60 - 30\sqrt{2}}{2} \approx 8.787$

$x < 8.79, \quad$ or $\quad x > 51.21$

Chapter Three Test

1. $a = -1$, $b = 2$, $c = 3$
    $2^2 - 4 \cdot (-1)(3) = 4 + 12 = 16$;
    2 real number solutions,
    2 x-intercepts.

2.

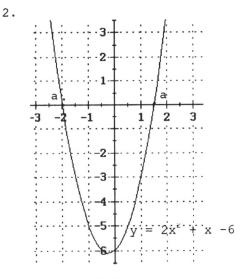

$y = 2x^2 + x - 6$

a) $\{-2, 1.5\}$
    $2x^2 + x - 6 = (x + 2)(2x - 3)$
    $(x + 2) = 0 \quad 2x - 3 = 0$
    $x = -2 \qquad\quad x = 1.5$

b) $x = -\dfrac{1}{4} \quad \left[-\dfrac{1}{2 \cdot 2} = -\dfrac{1}{4}\right]$

c) $\left(-\dfrac{1}{4},\ -6\dfrac{1}{8}\right)$

    $y = 2\left(-\dfrac{1}{4}\right)^2 + \left(-\dfrac{1}{4}\right) - 6$

    $= +\dfrac{1}{8} - \dfrac{2}{8} - 6 = -6\dfrac{1}{8}$

d) $x < -2$ or $x > 1.5$

e) One possible answer: $y = -10$

f) $-6\dfrac{1}{8} < y < -6$

3. $y = x^2 + 2$ is $y = x^2$ shifted up
    2 units. The vertex of
    $y = x^2 + 2$ is $(0, 2)$ and there
    are no x-intercepts.

4. $y = -x^2$ is $y = x^2$ reflected
    across the x-axis.

5. a) $\sqrt{-36} = \sqrt{-1} \cdot \sqrt{36} = 6i$

    b) $\sqrt{-75} = \sqrt{-1} \cdot \sqrt{25 \cdot 3} = 5i\sqrt{3}$

6. a) $\dfrac{5 + 10i}{5} = \dfrac{5}{5} + \dfrac{10i}{5} = 1 + 2i$

    b) $\dfrac{8 + 4i}{2} = \dfrac{8}{2} + \dfrac{4i}{2} = 4 + 2i$

7. a) $(1 + 5i)(1 - 5i)$
        $= 1 - 5i + 5i - 25i^2$
        $= 1 - 25i^2$
        $= 1 - 25(-1)$
        $= 1 + 25 = 26$

    b) $(5 - i)(5 + i)$
        $= 25 + 5i - 5i - i^2$
        $= 25 - i^2$
        $= 25 - (-1) = 26$

    c) $(1 - 5i)(1 - 5i)$
        $= 1 - 5i - 5i + 25i^2$
        $= 1 - 10i + (25)(-1)$
        $= 1 - 10i - 25$
        $= -24 - 10i$

8. $f(x) = (x - 1)(x^2 + x + 1) = 0$
    $x - 1 = 0$ or $x^2 + x + 1 = 0$
    $x = 1$

    $x = \dfrac{-1 \pm \sqrt{1^2 - 4 \cdot 1 \cdot 1}}{2 \cdot 1} = \dfrac{-1 \pm \sqrt{-3}}{2}$

    $x = -\dfrac{1}{2} \pm \dfrac{i\sqrt{3}}{2}$

    $\left\{1,\ -\dfrac{1}{2} \pm \dfrac{\sqrt{3}}{2}\,i\right\}$

9. $f(x) = x^2 - x - 2$

$(-1)^2 - 4 \cdot 1(-2)$

$= 1 + 8 = 9;$

2 x - intercepts

$x^2 - x - 2 = (x + 1)(x - 2)$

$\{-1, 2\} = x -$ intercepts;

$-\dfrac{b}{2a} = \dfrac{-(-1)}{2(1)} = \dfrac{1}{2}$

$y = \left(\dfrac{1}{2}\right)^2 - \left(\dfrac{1}{2}\right) - 2$

$= \dfrac{1}{4} - \dfrac{2}{4} - 2 = -2\dfrac{1}{4}$

Vertex $= \left(\dfrac{1}{2},\ -2\dfrac{1}{4}\right)$

10. Answers will vary.

11. $y = x^2 + 4x - 1$

$y + 1 = x^2 + 4x \qquad \left[\left(\dfrac{4}{2}\right)^2 = 4\right]$

$y + 1 + 4 = x^2 + 4x + 4$

$y + 5 = (x + 2)^2$

$y = (x + 2)^2 - 5$

12. Shift $y = x^2$ left 2 units and down 5 units.

13. $y = a(x - (-1))^2 + (-8) = a(x + 1)^2 - 8$

$19 = a(2 + 1)^2 - 8$

$19 = 9a - 8$

$27 = 9a \quad a = \dfrac{27}{9} = 3,$

$y = 3(x + 1)^2 - 8 = 3(x^2 + 2x + 1) - 8$

$y = 3x^2 + 6x - 5$

14. a) $y = a \cdot b$

b) $15 = 2a + b$

$b = 15 - 2a$

$y = a(15 - 2a) = 15a - 2a^2$

$y = -2a^2 + 15a$

c) Vertex a $= \dfrac{-15}{2(-2)} = \dfrac{15}{4}$

$y = -2\left(\dfrac{15}{4}\right)^2 + 15\left(\dfrac{15}{4}\right)$

$y = -2\left(\dfrac{15}{16}\right)^2 + \dfrac{15^2}{4}$

$= -\dfrac{15^2}{8} + \dfrac{2 \cdot 15^2}{8}$

$y = \dfrac{15^2}{8} = \dfrac{225}{8} = 28\dfrac{1}{8}$

Vertex $= \left(\dfrac{15}{4},\ 28\dfrac{1}{8}\right);$

Maximum area $= 28\dfrac{1}{8}$ sq ft.

CHAPTER FOUR

Section 4.0

1. a) $\dfrac{15}{45} = \dfrac{3 \cdot 5}{3 \cdot 3 \cdot 5} = \dfrac{1}{3}$

   Common factor $3 \cdot 5 = 15$

   b) $\dfrac{48}{160} = \dfrac{2 \cdot 2 \cdot 2 \cdot 2 \cdot 3}{2 \cdot 2 \cdot 2 \cdot 2 \cdot 2 \cdot 5} = \dfrac{3}{10}$

   Common factor $2 \cdot 2 \cdot 2 \cdot 2 = 16$

   c) $\dfrac{5280}{3600} = \dfrac{2 \cdot 2 \cdot 2 \cdot 2 \cdot 2 \cdot 3 \cdot 5 \cdot 11}{2 \cdot 2 \cdot 2 \cdot 2 \cdot 3 \cdot 3 \cdot 5 \cdot 5} = \dfrac{22}{15}$

   Common factor $2 \cdot 2 \cdot 2 \cdot 2 \cdot 3 \cdot 5 = 240$

3. a) $\dfrac{2\sqrt{3}}{4} = \dfrac{2 \cdot \sqrt{3}}{2 \cdot 2} = \dfrac{\sqrt{3}}{2}$

   Common factor 2

   b) $\dfrac{15\sqrt{2}}{5} = \dfrac{5 \cdot 3 \cdot \sqrt{2}}{5} = 3\sqrt{2}$

   Common factor 5

   c) $\dfrac{2 + 6\sqrt{2}}{6} = \dfrac{2\left(1 + 3\sqrt{2}\right)}{2 \cdot 3} = \dfrac{1 + 3\sqrt{2}}{3}$

   Common factor 2

5. a) $\dfrac{3}{5} = \dfrac{3 \cdot 7}{5 \cdot 7} = \dfrac{21}{35}$

   b) $\dfrac{5}{8} = \dfrac{5 \cdot 6}{8 \cdot 6} = \dfrac{30}{48}$

   c) $\dfrac{8}{3} = \dfrac{8 \cdot 16}{3 \cdot 16} = \dfrac{128}{48}$

7. $\dfrac{1}{4} + \dfrac{1}{10} = \dfrac{1 \cdot 5}{4 \cdot 5} + \dfrac{1 \cdot 2}{10 \cdot 2} = \dfrac{5}{20} + \dfrac{2}{20} = \dfrac{7}{20}$

   $\dfrac{1}{4} - \dfrac{1}{10} = \dfrac{5}{20} - \dfrac{2}{20} = \dfrac{3}{20}$

   $\dfrac{1}{4} \cdot \dfrac{1}{10} = \dfrac{1 \cdot 1}{4 \cdot 10} = \dfrac{1}{40}$

   $\dfrac{1}{4} \div \dfrac{1}{10} = \dfrac{1}{4} \cdot \dfrac{10}{1} = \dfrac{5 \cdot 2}{2 \cdot 2} = \dfrac{5}{2}$

9. $\dfrac{3}{4} + \dfrac{5}{6} = \dfrac{3 \cdot 3}{4 \cdot 3} + \dfrac{5 \cdot 2}{6 \cdot 2} = \dfrac{9}{12} + \dfrac{10}{12} = \dfrac{19}{12}$

   $\dfrac{3}{4} - \dfrac{5}{6} = \dfrac{9}{12} - \dfrac{10}{12} = -\dfrac{1}{12}$

   $\dfrac{3}{4} \cdot \dfrac{5}{6} = \dfrac{3 \cdot 5}{4 \cdot 6} = \dfrac{3 \cdot 5}{4 \cdot 3 \cdot 2} = \dfrac{5}{8}$

   $\dfrac{3}{4} \div \dfrac{5}{6} = \dfrac{3}{4} \cdot \dfrac{6}{5} = \dfrac{3 \cdot 3 \cdot 2}{2 \cdot 2 \cdot 5} = \dfrac{9}{10}$

11. $1\dfrac{1}{3} + 2\dfrac{1}{2} = \dfrac{4}{3} + \dfrac{5}{2} = \dfrac{4 \cdot 2}{3 \cdot 2} + \dfrac{5 \cdot 3}{2 \cdot 3}$

    $= \dfrac{8}{6} + \dfrac{15}{6} = \dfrac{23}{6}$

    $\dfrac{4}{3} - \dfrac{5}{2} = \dfrac{8}{6} - \dfrac{15}{6} = -\dfrac{7}{6}$

    $\dfrac{4}{3} \cdot \dfrac{5}{2} = \dfrac{4 \cdot 5}{3 \cdot 2} = \dfrac{20}{6} = \dfrac{10}{3}$

    $\dfrac{4}{3} \div \dfrac{5}{2} = \dfrac{4}{3} \cdot \dfrac{2}{5} = \dfrac{4 \cdot 2}{3 \cdot 5} = \dfrac{8}{15}$

13. $\dfrac{186\ cm^2}{6\ cm} = \dfrac{3 \cdot 2 \cdot 31\ cm \cdot cm}{3 \cdot 2\ cm} = 31\ cm$

15. $\dfrac{36\ in}{1728\ in^3} = \dfrac{36 \cdot in}{36 \cdot 48 \cdot in \cdot in \cdot in} = \dfrac{1}{48\ in^2}$

17. $\dfrac{1500\ foot \cdot pounds}{25\ ft} = \dfrac{25 \cdot 60\ foot \cdot pounds}{25\ ft}$

    $= 60\ pounds$

19. a) $1\ gram \cdot \dfrac{9\ calories}{gram} = 9\ calories$

    b) $41\ grams \cdot \dfrac{4\ calories}{gram} = 164\ calories$

    c) $5\ grams \cdot \dfrac{4\ calories}{gram} = 20\ calories$

    $9 + 164 + 20 = 193\ calories$

21. a) $8 \text{ grams} \cdot \dfrac{9 \text{ calories}}{\text{gram}} = 72 \text{ calories}$

   b) $27 \text{ grams} \cdot \dfrac{4 \text{ calories}}{\text{gram}} = 108 \text{ calories}$

   c) $4 \text{ grams} \cdot \dfrac{4 \text{ calories}}{\text{gram}} = 16 \text{ calories}$

   $72 + 108 + 16 = 196 \text{ calories}$

23. a) $d = \dfrac{1}{2} \cdot 32.2 \dfrac{ft}{\sec^2} \cdot (1 \sec)^2$

   $= \dfrac{1}{2} \cdot 32.2 \cdot 1 \cdot \dfrac{ft}{\sec^2} \cdot \sec^2$

   $d = 16.1 \text{ ft}$

   b) $d = \dfrac{1}{2} \cdot 32.2 \dfrac{ft}{\sec^2} \cdot (3 \sec)^2$

   $= \dfrac{1}{2} \cdot 32.2 \cdot 9 \cdot \dfrac{ft}{\sec^2} \cdot \sec^2$

   $d = 144.9 \text{ ft}$

   c) $d = \dfrac{1}{2} \cdot 32.2 \dfrac{ft}{\sec^2} \cdot (9 \sec)^2$

   $= \dfrac{1}{2} \cdot 32.2 \cdot 81 \cdot \dfrac{ft}{\sec^2} \cdot \sec^2$

   $d = 1304.1 \text{ ft}$

25. a) $S = 9.81 \dfrac{m}{\sec^2} \cdot 1 \sec = 9.81 \dfrac{m}{\sec}$

   b) $S = 9.81 \dfrac{m}{\sec^2} \cdot 2 \sec = 19.62 \dfrac{m}{\sec}$

   c) $S = 9.81 \dfrac{m}{\sec^2} \cdot 4 \sec = 39.24 \dfrac{m}{\sec}$

27. $300 \text{ mL} \cdot \dfrac{1 \text{ liter}}{1000 \text{ mL}} = 0.3 \text{ liter}$

29. $25 \text{ ft} \cdot \dfrac{12 \text{ in}}{1 \text{ ft}} \cdot \dfrac{1 \text{ m}}{39.37 \text{ in}} \approx 7.620 \text{ m}$

31. $150 \text{ ft}^3 \cdot \left(\dfrac{1 \text{ yd}}{3 \text{ ft}}\right)^3 = 150 \text{ ft}^3 \cdot \dfrac{1 \text{ yd}^3}{27 \text{ ft}^3}$

   $\approx 5.556 \text{ yd}^3$

33. $100 \text{ in}^3 \cdot \left(\dfrac{1 \text{ ft}}{12 \text{ in}}\right)^3 = 100 \text{ in}^3 \cdot \dfrac{1 \text{ ft}^3}{1728 \text{ in}^3}$

   $\approx 0.058 \text{ ft}^3$

35. $200 \text{ mL} \cdot \dfrac{1 \text{ kg}}{1000 \text{ mL}} \cdot \dfrac{1000 \text{ g}}{1 \text{ kg}} = 200 \text{ g}$

37. $\dfrac{40 \text{ miles}}{1 \text{ hr}} \cdot \dfrac{5280 \text{ ft}}{1 \text{ mile}} \cdot \dfrac{1 \text{ hr}}{60 \text{ min}} \cdot \dfrac{1 \text{ min}}{60 \text{ sec}} \approx 58.667 \dfrac{ft}{\sec}$

39. $\dfrac{240 \text{ mL}}{12 \text{ hr}} \cdot \dfrac{60 \text{ microdrops}}{1 \text{ mL}} \cdot \dfrac{1 \text{ hr}}{60 \text{ min}} = 20 \dfrac{\text{microdrops}}{\text{min}}$

41. a) $\dfrac{3}{5} \boxdot \dfrac{2}{7} = \dfrac{6}{35}$, Multiply

   b) $\dfrac{3}{5} \boxplus \dfrac{2}{7} = \dfrac{3}{5} \cdot \dfrac{7}{2} = \dfrac{21}{10}$, Divide

43. a) $\dfrac{3}{5} \boxplus \dfrac{2}{7} = \dfrac{3 \cdot 7}{5 \cdot 7} + \dfrac{2 \cdot 5}{7 \cdot 5}$

   $= \dfrac{21}{35} + \dfrac{10}{35} = \dfrac{31}{35}$, Add

   b) $\dfrac{5}{7} \boxminus \dfrac{2}{3} = \dfrac{5 \cdot 3}{7 \cdot 3} - \dfrac{2 \cdot 7}{3 \cdot 7}$

   $= \dfrac{15}{21} - \dfrac{14}{21} = \dfrac{1}{21}$, Subtract

45. Wrong. Multiply numerator and denominator by the same number when setting up the common denominator.

   $\dfrac{3}{4} + \dfrac{2}{3} = \dfrac{3 \cdot 3}{4 \cdot 3} + \dfrac{2 \cdot 4}{3 \cdot 4} = \dfrac{9}{12} + \dfrac{8}{12}$

   $= \dfrac{9 + 8}{12} = \dfrac{17}{12}$

47. Right, Method always works. Disadvantage: may result in fractional expression in numerator and/or denominator.

49. Wrong, find a common denominator before adding.

   $\dfrac{5}{6} + \dfrac{1}{3} = \dfrac{5}{6} + \dfrac{1 \cdot 2}{3 \cdot 2} = \dfrac{5}{6} + \dfrac{2}{6} = \dfrac{7}{6}$

Section 4.1

1. a) $\dfrac{\frac{1}{2}\text{ foot}}{2\text{ inches}} = \dfrac{\frac{1}{2}\text{ foot} \cdot \frac{12\text{ in}}{1\text{ foot}}}{2\text{ in}} = \dfrac{6\text{ in}}{2\text{ in}} = \dfrac{3}{1}$

   b) $\dfrac{3000\text{ grams}}{6\text{ kilograms}} = \dfrac{3000 \cdot \frac{1\text{ kg}}{1000\text{ g}}}{6\text{ kg}} = \dfrac{3\text{ kg}}{6\text{ kg}} = \dfrac{1}{2}$

   c) $\dfrac{32\text{ ounces}}{6\text{ pounds}} = \dfrac{32\text{ oz} \cdot \frac{1\text{ lb}}{16\text{ oz}}}{6\text{ lb}} = \dfrac{2\text{ lb}}{6\text{ lb}} = \dfrac{1}{3}$

3. a) $\dfrac{300\text{ mL}}{30\text{ L}} = \dfrac{300\text{ mL}}{30\text{ L} \cdot \frac{1000\text{ mL}}{\text{L}}}$

   $= \dfrac{300\text{ mL}}{30000\text{ mL}} = \dfrac{1}{100}$

   b) $\dfrac{2\text{ yr}}{180\text{ month}} = \dfrac{2\text{ yr}}{180\text{ month} \cdot \frac{1\text{ yr}}{12\text{ month}}}$

   $= \dfrac{2\text{ yr}}{15\text{ yr}} = \dfrac{2}{15}$

   c) $\dfrac{40\text{ minutes}}{\frac{1}{4}\text{ hour}} = \dfrac{40\text{ min}}{\frac{1}{4}\text{ hr} \cdot \frac{60\text{ min}}{1\text{ hr}}}$

   $= \dfrac{40\text{ min}}{\frac{60}{4}\text{ min}} = \dfrac{40}{15} = \dfrac{8}{3}$

5. a) $\dfrac{4.5\text{ in}}{1.68\text{ in}} = \dfrac{4.5 \cdot 100}{1.68 \cdot 100} = \dfrac{450}{168}$

   $= \dfrac{75 \cdot 6}{28 \cdot 6} = \dfrac{75}{28} \approx 2.679\text{ to }1$

   b) Area hole $= \dfrac{\pi \cdot 4.5^2}{4}$

   Area ball $= \dfrac{\pi \cdot 1.68^2}{4}$

   $\dfrac{\frac{\pi \cdot 4.5^2}{4}}{\frac{\pi \cdot 1.68^2}{4}} = \dfrac{4.5^2}{1.68^2} = \dfrac{4.5}{1.68} \cdot \dfrac{4.5}{1.68}$

   Using results from a (above)

   $\dfrac{75}{28} \cdot \dfrac{75}{28} = \dfrac{5625}{784} \approx 7.175\text{ to }1$

7. a) $A = \dfrac{\pi \cdot (1.66\text{mm})^2}{4} = \dfrac{\pi \cdot 2.56}{4}\text{ mm}^2$

   $\approx 2.011\text{mm}^2$

   b) $\dfrac{1\text{ cm}^2}{2.011\text{ mm}^2} = \dfrac{1\text{ cm}^2 \cdot \left(\frac{10\text{ mm}}{1\text{ cm}}\right)^2}{2.011\text{ mm}^2}$

   $= \dfrac{100\text{ mm}^2}{2.011\text{ mm}^2} = \dfrac{100}{2.011} \approx 49.727\text{ to }1$

9. $\dfrac{a}{b} \cdot bd = \dfrac{c}{d} \cdot bd \Rightarrow ad = cb$

11. $\dfrac{3}{x} = \dfrac{5}{14}$, $3 \cdot 14 = 5 \cdot x$, $42 = 5x$, $x = 8.4$

13. $\dfrac{x + 1}{6} = \dfrac{x}{9}$, $9(x + 1) = 6x$, $9x + 9 = 6x$,

    $3x = -9$, $x = -3$

15. $\dfrac{x - 7}{10} = \dfrac{-3}{x + 6}$, $(x - 7)(x + 6) = -30$,

    $x^2 - x - 42 = -30$, $x^2 - x - 12 = 0$,

    $(x - 4)(x + 3) = 0$

    $\{-3, 4\}$

17. $\dfrac{6}{4} = \dfrac{f}{8}$, $48 = 4f$, $f = 12$

    $\dfrac{6}{4} = \dfrac{g}{4\sqrt{3}}$, $24\sqrt{3} = 4g$, $g = 6\sqrt{3}$

19. $\dfrac{4\sqrt{3}}{10} = \dfrac{8}{f}$, $4\sqrt{3}f = 80$, $f = \dfrac{80}{4\sqrt{3}}$

    $f = \dfrac{20}{\sqrt{3}} = \dfrac{20\sqrt{3}}{3}$

    $\dfrac{4\sqrt{3}}{10} = \dfrac{4}{e}$, $4\sqrt{3}e = 40$,

    $e = \dfrac{40}{4\sqrt{3}} = \dfrac{10}{\sqrt{3}} = \dfrac{10\sqrt{3}}{3}$

21. $\dfrac{\sqrt{5}}{4} = \dfrac{f}{8}$, $8\sqrt{5} = 4f$, $f = 2\sqrt{5}$

    $\dfrac{\sqrt{5}}{4} = \dfrac{g}{4\sqrt{3}}$, $4\sqrt{15} = 4g$, $g = \sqrt{15}$

Section 4.1

23. $2^2 + 2^2 = n^2$, $8 = n^2$, $n = \sqrt{8} = 2\sqrt{2}$

$\dfrac{2}{x} = \dfrac{2\sqrt{2}}{8}$, $16 = 2\sqrt{2}x$, $x = \dfrac{16}{2\sqrt{2}}$

$x = \dfrac{8}{\sqrt{2}} = \dfrac{8\sqrt{2}}{2} = 4\sqrt{2}$

x & $y = 4\sqrt{2}$

25.

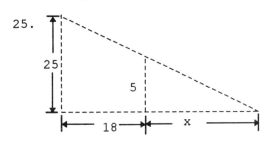

$\dfrac{5}{25} = \dfrac{x}{x+18}$, $5(x + 18) = 25x$,

$5x + 90 = 25x$, $90 = 20x$,

$x = \dfrac{90}{20} = 4.5$ ft

27.

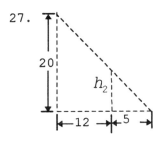

$\dfrac{20}{h_2} = \dfrac{17}{5}$, $100 = 17h_2$ $h_2 = \dfrac{100}{17} \approx 5.9\,\text{ft}$

29. $\dfrac{8\text{ft} \cdot 12\text{ft} \cdot 4\text{ft}}{3 \cdot 4\text{ft}^3 / \text{bale}} = \dfrac{8 \cdot 12}{3}\dfrac{\text{ft}^3}{\dfrac{\text{ft}^3}{\text{bale}}}$

$= 32$ bales

31. a) Input x, weight in pounds; Output y, distance stretched in inches.

$m = \dfrac{6 - 4}{15 - 10} = \dfrac{2}{5}$;

$4 = \dfrac{2}{5}(10) + b$, $b = 0$

$y = \dfrac{2}{5}x$

b) Proportional

33. a) Input x, no. of credit hours; output y, cost in $.

$m = \dfrac{345 - 185(\$)}{4 - 2\,\text{hr}} = \dfrac{160}{2} = \dfrac{\$80}{\text{cr.hr.}}$;

$\$345 = \$80(4) + b$, $b = \$25$

$y = \$80x + \$25$

b) Not proportional

c) y-intercept represents student fees.

35. a) Input x, purchase price in $; output y, tax in $.

$m = \dfrac{2.25 - 1.20}{30 - 16} = \dfrac{1.05}{14} = 0.075$;

$\$2.25 = 0.075(30) + b$, $b = 0$

$y = 0.075x$

b) Proportional

37. $k = \dfrac{110 - 55}{2 - 1}\dfrac{\text{miles}}{\text{hours}} = 55\dfrac{\text{miles}}{\text{hour}}$

$55\dfrac{\text{miles}}{\text{hour}} \cdot 8\,\text{hr.} = 55 \cdot 8\,\text{miles}$

$= 440$ miles

39. $k = \dfrac{\$86.94 - \$57.96}{6 - 4 \text{ cd's}}$

    $= \dfrac{\$28.98}{2 \text{ cd's}} = \dfrac{\$14.49}{1 \text{ cd}}$

    $\dfrac{\$14.49}{\text{cd}} \cdot 10 \text{ cd} = \$144.90$

41. 7 inch diameter $\Rightarrow$ 3.5 inch radius
    3 inch diameter $\Rightarrow$ 1.5 inch radius
    weight $= k(\text{radius})^2$

    $k = \dfrac{4.5 oz}{(3.5 \text{ in})^2} = \dfrac{4.5 oz}{12.25 \text{ in}^2} \approx 0.367 \dfrac{oz}{\text{in}^2}$

    weight $= 0.367 \dfrac{oz}{\text{in}^2} \cdot (1.5 \text{ in})^2$

    ≈ 0.826 oz.

43. 14a) $y = 0.49x$

    14b) $m = \dfrac{4.44 - 2.10}{14 - 5} = \dfrac{2.34}{9} = 0.26;$

    $4.44 = 0.26(14) + b, \quad b = 0.80$

    $y = 0.26x + 0.80,$
    y-intercept is charge to connect to long distance destination.

49. a) $A = \dfrac{1}{2}\text{bh}, \quad k = \dfrac{1}{2}$

    b) $A = \text{LW}, \quad k = 1$

    c) $V = \dfrac{1}{3}\pi r^2 h, \quad k = \dfrac{1}{3}\pi$

51. $V = \dfrac{1}{3}Bh, \quad V = \dfrac{1}{3}(53095m^2)(146.6m)$

    $\approx 2{,}594{,}576 m^3$

53. a) $c = \text{kr}$

    b) $k = 2\pi, (c = 2\pi r)$

55. a) $A = kr^2$

    b) $k = \pi, (A = \pi r^2)$

57. a) $D = \text{krt}$

    b) $k = 1, (D = rt)$

59. a) $A = ks^2$

    b) $k = 6, (A = 6s^2)$

61. Needed to use the reciprocal of 4 quarts to 1 gallon.

    $\dfrac{3 \text{ quarts}}{2 \text{ gallons}} \cdot \dfrac{1 \text{ gallon}}{4 \text{ quarts}} = \dfrac{3}{8}$

63. Cross multiplication gives an equation, not a fraction.

    $\dfrac{4}{15} = \dfrac{6}{x}, \quad 4x = 90, \quad x = 22.5$

Section 4.2

1. $y = yrs$, $x = \dfrac{yd^3}{yr}$, $y = \dfrac{1{,}000{,}000}{x}$

   $k = 1{,}000{,}000\ yd^3$

3. $y = yrs$, $x = \dfrac{\text{metric tons}}{yr}$

   $y = \dfrac{5.74 \times 10^8}{x}$

   $k = 5.74 \times 10^8$ metric tons

5. $y = \dfrac{\text{points}}{\text{problem}}$, $x = $ no. of problems

   $y = \dfrac{100}{x}$

   $k = 100$ points

7. $35\,\text{lbs} \cdot 6\,ft = x\,\text{lbs} \cdot 4\,ft$

   $x = 52.5\,\text{lbs}$, $k = 210\,\text{lb ft}$

9. 80 servings $\cdot\ 0.75\ \dfrac{\text{cup}}{\text{serving}}$

   $= x$ servings $\cdot\ 1\ \dfrac{\text{cup}}{\text{serving}}$

   $x = 60$ servings, $k = 60$ cups

11. 133 years $\cdot\ 900$ Thousand metric tons

    $= x$ years $\cdot\ 1300$ Thousand metric tons

    $x \approx 92.077$ years

    $k = 119{,}700$ Thousand metric tons

13. $4\,\text{in} \cdot 900\ \dfrac{ft}{gal} = 3\,\text{in} \cdot x\ \dfrac{ft}{gal}$

    $x = 1200\ \dfrac{ft}{gal}$, $k = 3600\ \dfrac{\text{in} \cdot ft}{gal}$

    Alternate answer:

    $\dfrac{1}{3}\,\text{ft} \cdot 900\ \dfrac{ft}{gal} = \dfrac{1}{4}\,\text{ft} \cdot x\ \dfrac{ft}{gal}$

    $x = 1200\ \dfrac{ft}{gal}$, $k = 300\ \dfrac{ft^2}{gal}$

15. Inverse variation,

    $k = 5600\ \dfrac{\text{tons} \cdot \text{yrs}}{\text{day}}$

    or $k = 5600\ \dfrac{\text{tons} \cdot \text{yrs}}{\text{day}} \cdot \dfrac{365\ \text{days}}{\text{yr}}$

    $= 2{,}044{,}000$ tons

17. Direct variation,

    $k = \$0.21875\ \dfrac{\text{cost}}{\text{item}}$

19. Inverse variation, $k = 90$ miles

21. Inverse variation,

    $k = 12$ worker $\cdot$ days

23. Direct variation, $k = \$1100$ per semester

25. $y_1 = \dfrac{k}{3^2}$, $y_2 = \dfrac{k}{6^2}$; $9y_1 = k$, $36y_2 = k$;

    $9y_1 = 36y_2$, $y_1 = 4y_2$, 4 times greater

27. $\dfrac{x_1 y_1}{x_2 y_1} = \dfrac{x_2 y_2}{x_2 y_1}$, $\dfrac{x_1}{x_2} = \dfrac{y_2}{y_1}$

    $\dfrac{y_1}{y_2}$ has been inverted

29. $t = \dfrac{6}{R}$

Rate (mph)	time (hrs)
1	6
2	3
3	2
4	1.5
5	1.2
6	1
7	0.857
8	0.75
9	0.667
10	0.6

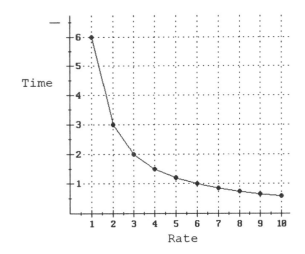

31. a)

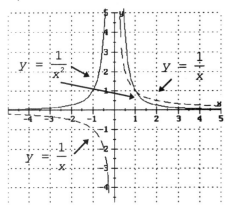

b) $x > 1$

c) $x = 1$

d) $\dfrac{1}{x^2}$ is positive for $x < 0$,

$\dfrac{1}{x}$ is negative for $x < 0$.

33. a)

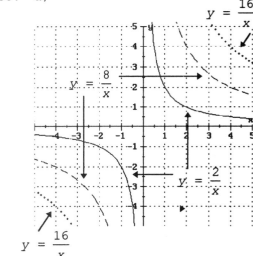

Section 4.2

33. b) All are in first and third
quadrants.

   c) $\dfrac{8}{x}$ is flatter than $\dfrac{2}{x}$, $\dfrac{16}{x}$ is
   flatter than $\dfrac{8}{x}$; in the first
   quadrant $\dfrac{16}{x}$ is above $\dfrac{8}{x}$ and
   $\dfrac{2}{x}$. This is reversed in the
   third quadrant.

   d) k = the numerators 2, 8, 16.

   e) Changes its distance from the
   x-axis by a constant factor.

   f) Below in first quadrant and
   above in the third quadrant.

   g) Above in first quadrant and
   below in the third quadrant.

35. a) $\dfrac{20\,\text{miles}}{65\,\dfrac{\text{miles}}{\text{hour}}} \approx 0.3077\,\text{hr}$,

   $0.3077\,\text{hr} \cdot \dfrac{60\,\text{min.}}{1\,\text{hr.}}$

   $\approx 18.462\,\text{min.}$

   b)

Mph	Min
60	20
65	18.462
70	17.143
75	16
80	15

   c) 18.462−5 = 13.462 min.

   $13.462\,\text{min} \cdot \dfrac{1\,\text{hr}}{60\,\text{min}} \approx 0.2244\,\text{hrs.}$

   $\dfrac{20\,\text{miles}}{x\,\dfrac{\text{miles}}{\text{hr}}} = 0.2244\,\text{hrs,}$

   $x \approx 89.127\ \text{MPH}$

37. a) $\dfrac{120v}{12v} = \dfrac{100mA}{I_1}$,

   $I_1 = \dfrac{12v \cdot 100mA}{120v} = 10mA$

   b) $\dfrac{V_1}{V_2} = \dfrac{I_2}{I_1}$, $V_1 I_1 = V_2 I_2$,
   Inversely proportional

   c) $k = 12v \cdot 100 \times 10^{-3} A$
   = 1.2 Volt Amp

Section 4.3

1. $x + 3 = 0,\quad x = -3$

3. $4 - a = 0,\quad a = 4$

5. $x^2 - 3x + 28 = 0$

$$\frac{-(-3) \pm \sqrt{(-3)^2 - 4(1)\,(28)}}{2 \cdot 1}$$

$$\frac{3 \pm \sqrt{-103}}{2}$$

Determinent is negative.
No real number solutions.

Expression is defined for all real numbers.

7. $\dfrac{-2}{6} = \dfrac{-1 \cdot 2}{3 \cdot 2} = \dfrac{-1}{3} = -\dfrac{1}{3}$

$\dfrac{2}{-6} = \dfrac{2 \cdot 1}{2 \cdot (-3)} = \dfrac{1}{-3} = -\dfrac{1}{3}$

9. $\dfrac{-6}{-4} = \dfrac{-1 \cdot 2 \cdot 3}{-1 \cdot 2 \cdot 2} = \dfrac{3}{2}$

$-\dfrac{6}{4} = -1 \cdot \dfrac{2 \cdot 3}{2 \cdot 2} = -1 \cdot \dfrac{3}{2} = -\dfrac{3}{2}$

11. $\dfrac{-a}{b} = \dfrac{-1 \cdot a}{b} = -1 \cdot \dfrac{a}{b} = -\dfrac{a}{b}$

$-\dfrac{a}{b} = \dfrac{1 \cdot a}{-1 \cdot b} = \dfrac{1}{-1} \cdot \dfrac{a}{b}$

$= -1 \cdot \dfrac{a}{b} = -\dfrac{a}{b}$

13. $\dfrac{6 - 4}{4 - 6} = \dfrac{2}{-2} = -1$

Numerator is the additive inverse of the denominator.

15. $\dfrac{-3 - 7}{7 - (-3)} = \dfrac{-10}{10} = -1$

17. a) $-1(x + y) = -x - y$

b) $-1(-x + y) = x - y$

c) $-1(y - x) = -y + x$ or $x - y$

19. a) $\dfrac{x - 3}{\Box} = 1,\quad \dfrac{x - 3}{1} = \Box,$

$\Box = x - 3,\quad x \neq 3$

b) $\dfrac{x + 2}{\Box} = -1,\quad \dfrac{x + 2}{-1} = \Box,$

$-1(x + 2) = \Box,\quad \Box = -x - 2,$

$x \neq -2$

c) $\dfrac{b - a}{\Box} = 1,\quad \dfrac{b - a}{1} = \Box,$

$\Box = b - a,\quad b \neq a$

d) $\dfrac{3 - x}{\Box} = -1,\quad \dfrac{3 - x}{-1} = \Box,$

$-1(3 - x) = \Box$

$\Box = -3 + x = x - 3,$

$x \neq 3$

21. a) $\dfrac{16a^2}{10a} = \dfrac{8 \cdot 2 \cdot \overset{\checkmark}{a} \cdot \overset{\checkmark}{a}}{5 \cdot \underset{\checkmark}{2} \cdot \underset{\checkmark}{a}} = \dfrac{8a}{5},\ a \neq 0$

b) $\dfrac{2x^2y}{10xy^2} = \dfrac{\overset{\checkmark}{2} \cdot \overset{\checkmark}{x} \cdot x \cdot \overset{\checkmark}{y}}{5 \cdot \underset{\checkmark}{2} \cdot \underset{\checkmark}{x} \cdot \underset{\checkmark}{y} \cdot y}$

$= \dfrac{x}{5y},\quad x \neq 0 \text{ and } y \neq 0$

23. a) $\dfrac{xy}{2x + y}$ Already simplified

$2x \neq -y$

b) $\dfrac{xy}{2x + xy} = \dfrac{x \cdot y}{x(2 + y)}$

$= \dfrac{y}{2 + y},\quad x \neq 0 \text{ and } y \neq -2$

83

25. a) $\dfrac{3 - x}{(x + 3)(x - 3)} = \dfrac{-1(-3 + x)}{(x + 3)(x - 3)}$

$= \dfrac{-1(x - 3)^{\checkmark}}{(x + 3)(x - 3)_{\checkmark}} = \dfrac{-1}{x + 3}, \quad x \neq -3, 3$

b) $\dfrac{x^2 - 9}{x^2 + 5x + 6} = \dfrac{(x + 3)^{\checkmark}(x - 3)}{(x + 3)_{\checkmark}(x + 2)}$

$= \dfrac{x - 3}{x + 2}, \quad x \neq -3, -2$

27. a) $\dfrac{2ac + 4bc}{4ad + 8bd} = \dfrac{2c(a + 2b)}{4d(a + 2b)}$

$= \dfrac{c}{2d}, \quad a \neq -2b, \ d \neq 0$

b) $\dfrac{6x^2 + 3x}{12x^2 - 6x} = \dfrac{3x(2x + 1)}{6x(2x - 1)}$

$= \dfrac{2x + 1}{2(2x - 1)}, \quad x \neq 0, \dfrac{1}{2}$

29. a) $\dfrac{x^2 + x - 6}{2 - x} = \dfrac{(x + 3)(x - 2)}{-1(x - 2)} = \dfrac{x + 3}{-1}$

$= -(x + 3), \quad x \neq 2$

b)

$\dfrac{x - 3}{6 - 2x} = \dfrac{x - 3}{-2(x - 3)} = -\dfrac{1}{2}, \quad x \neq 3$

31. a) $\dfrac{1}{a} \div \dfrac{1}{b} = \dfrac{\dfrac{1}{a}}{\dfrac{1}{b}} = \dfrac{1}{a} \cdot \dfrac{b}{1} = \dfrac{b}{a}$

b) $\dfrac{1}{b} \cdot \dfrac{1}{a} = \dfrac{1}{ba} = \dfrac{1}{ab}$

c) $\dfrac{1}{b} \div a = \dfrac{\dfrac{1}{b}}{a} = \dfrac{1}{b} \cdot \dfrac{1}{a} = \dfrac{1}{ba} = \dfrac{1}{ab}$

d) $\dfrac{1}{b} \div b = \dfrac{\dfrac{1}{b}}{b} = \dfrac{1}{b} \cdot \dfrac{1}{b} = \dfrac{1}{b^2}$

e) $\dfrac{a}{b} \div \dfrac{1}{b} = \dfrac{\dfrac{a}{b}}{\dfrac{1}{b}} = \dfrac{a}{b} \cdot \dfrac{b}{1} = a$

f) $b \cdot \dfrac{1}{b} = \dfrac{b}{1} \cdot \dfrac{1}{b} = 1$

33. a) $\dfrac{1}{x} \cdot \dfrac{x^2}{1} = \dfrac{x \cdot x}{x} = x$

b) $\dfrac{1}{a} \div \dfrac{a^2 b^2}{1} = \dfrac{1}{a} \cdot \dfrac{1}{a^2 b^2} = \dfrac{1}{a^3 b^2}$

c) $\dfrac{b}{a} \div \dfrac{a^2}{b^2} = \dfrac{b}{a} \cdot \dfrac{b^2}{a^2} = \dfrac{b^3}{a^3}$

d) $\dfrac{x}{y} \div \dfrac{x^3}{y^2} = \dfrac{x}{y} \cdot \dfrac{y^2}{x^3} = \dfrac{x \cdot \overset{\checkmark}{y} \cdot \overset{\checkmark}{y}}{y \cdot x \cdot x \cdot x} = \dfrac{y}{x^2}$

35. a) $\dfrac{a^2 + 7a + 12}{a^2 - 4} \div \dfrac{a^2 + 4a}{a - 2}$

$= \dfrac{(a + 4)(a + 3)}{(a + 2)(a - 2)} \cdot \dfrac{(a - 2)}{a(a + 4)}$

$= \dfrac{a + 3}{a(a + 2)}$

b) $\dfrac{x^2 - 2x}{x} \cdot \dfrac{x^2 - 1}{x^2 - 3x + 2}$

$= \dfrac{x(x - 2)}{x} \cdot \dfrac{(x + 1)(x - 1)}{(x - 2)(x - 1)}$

$= x + 1$

c) $\dfrac{b - 3}{b^2 - 4b + 3} \div \dfrac{b^2 - b}{b - 1}$

$= \dfrac{(b - 3)}{(b - 3)(b - 1)} \cdot \dfrac{(b - 1)}{b(b - 1)}$

$= \dfrac{1}{b(b - 1)}$

35. d) $\dfrac{x^2 - 6x + 9}{x^2 + 3x} \div \dfrac{x - 3}{x + 3}$

$$= \dfrac{(x - 3)(x - 3)}{x(x + 3)} \cdot \dfrac{x + 3}{x - 3}$$

$$= \dfrac{x - 3}{x}$$

e) $\dfrac{x^2 + 3x}{x} \cdot \dfrac{x^2 - x - 6}{x^2 - 9}$

$$= \dfrac{x(x + 3)}{x} \cdot \dfrac{(x - 3)\,(x + 2)}{(x - 3)\,(x + 3)}$$

$$= x + 2$$

f) $\dfrac{4a^2 + 4a + 1}{4 - 9a^2} \div \dfrac{4a^2 + 2a}{3a - 2}$

$$= \dfrac{(2a + 1)\,(2a + 1)}{(2 - 3a)\,(2 + 3a)} \cdot \dfrac{3a - 2}{2a(2a + 1)}$$

$$= \dfrac{(2a + 1)}{(2 - 3a)\,(2 + 3a)} \cdot \dfrac{-1(2 - 3a)}{2a}$$

$$= \dfrac{-(2a + 1)}{2a(2 + 3a)} = \dfrac{-2a - 1}{2a(2 + 3a)}$$

37. $\dfrac{2}{3} + \dfrac{4}{7} = \dfrac{2 \cdot 7}{3 \cdot 7} + \dfrac{4 \cdot 3}{7 \cdot 3}$

$$= \dfrac{14}{21} + \dfrac{12}{21} = \dfrac{26}{21}$$

Reciprocal $= \dfrac{21}{26}$

$\dfrac{3}{2} + \dfrac{7}{4} = \dfrac{6}{4} + \dfrac{7}{4} = \dfrac{13}{4}$

39. $\dfrac{a}{b} + \dfrac{c}{d} = \dfrac{a \cdot d}{b \cdot d} + \dfrac{c \cdot b}{d \cdot b} = \dfrac{ad + cb}{bd}$

Reciprocal $= \dfrac{bd}{ad + cb}$

41. Correct procedure is:

$\dfrac{x^2 + x + 2}{x}$ numerator does not factor, this is already simplified.

43. Correct procedure is:

$\dfrac{x^2 - 3x + 2}{2 - x} = \dfrac{(x - 2)\,(x - 1)}{-1(x - 2)}$

$= -(x - 1) = -x + 1$

45. Correct procedure is:

$\dfrac{x^2 - 2x + 1}{x - 2} = \dfrac{(x - 1)\,(x - 1)}{x - 2}$

No common factor, already simplified.

47. Multiplication is not distributive over multiplication.

49. $\dfrac{\frac{50 \text{ miles}}{1}}{\frac{10 \text{ miles}}{\text{hr}}} = \dfrac{50 \text{ miles}}{1} \cdot \dfrac{\text{hr}}{10 \text{ miles}} = 5 \text{ hr}$

51. $\dfrac{\frac{5280 \text{ ft}}{88 \text{ ft}}}{\text{sec}} = \dfrac{5280 \text{ ft}}{1} \cdot \dfrac{\text{sec}}{88 \text{ ft}} = 60 \text{ sec}$

53. $\dfrac{\frac{65 \text{ miles}}{\text{hr}}}{\frac{15 \text{ miles}}{\text{gal}}} = \dfrac{65 \text{ miles}}{\text{hr}} \cdot \dfrac{\text{gal}}{15 \text{ miles}}$

$= \dfrac{13}{3} \dfrac{\text{gal}}{\text{hr}} = 4\dfrac{1}{3} \text{ gal per hr}$

55. $\dfrac{\frac{24 \text{ cans}}{\text{case}}}{\frac{\$3.98}{\text{case}}} = \dfrac{24 \text{ cans}}{\text{case}} \cdot \dfrac{\text{case}}{\$3.98}$

$\approx 6.030 \text{ cans per dollar}$

57. $\dfrac{\frac{8 \text{ stitches}}{\text{inch}}}{\frac{1 \text{ ft}}{12 \text{ inches}}} = \dfrac{8 \text{ stitches}}{\text{inch}} \cdot \dfrac{12 \text{ inches}}{1 \text{ ft}}$

$= 96 \text{ stitches per foot}$

59. $\dfrac{\frac{95 \text{ words}}{\text{min}}}{\frac{300 \text{ words}}{\text{page}}} = \dfrac{95 \text{ words}}{\text{min}} \cdot \dfrac{\text{page}}{300 \text{ words}}$

$\approx 0.317 \text{ pages per minute}$

61. $\dfrac{\frac{220 \text{ days}}{5 \text{ days}}}{\text{week}} = \dfrac{220 \text{ days}}{1} \cdot \dfrac{\text{week}}{5 \text{ days}}$

$= 44 \text{ weeks}$

Section 4.3

63. a) Yes

     b) 0 in denominator

     c) 20; ($y = 20 + x$ and
        $y = 2x$)

     d) 1

     e) $y_3 = \dfrac{x + 20}{x} = \dfrac{x}{x} + \dfrac{20}{x} = 1 + \dfrac{20}{x}$,

        as x gets large $\dfrac{20}{x}$ gets small.

     h) Increases rapidly

     i) Function is undefined at
        $x = 0$.

65.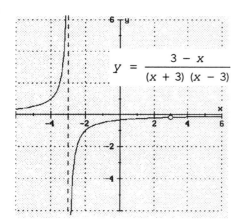

     e) Hole before simplifying at
        $x = 3$.

     f) Nearly vertical around
        $x = -3$.

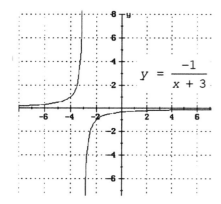

67.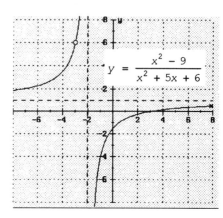

     e) Hole before simplifying at
        $x = -3$

     f) Nearly vertical around
        $x = -2$.

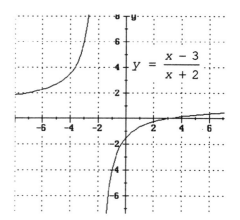

86

69.

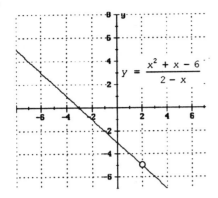

c) Straight line

d) Slope, −1

e) Hole before simplifying at
   x = 2.

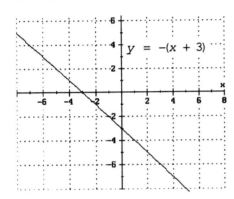

Mid Chapter 4 Test

1. $\dfrac{32}{24} = \dfrac{8 \cdot 4}{8 \cdot 3} = \dfrac{4}{3}$

2. $\dfrac{3}{3(1 + \sqrt{3})} = \dfrac{1}{1 + \sqrt{3}}$

3. $\dfrac{a\,b\,c}{a\,c\,es} = \dfrac{b}{es}$

4. $\dfrac{2ab}{6a^2b} = \dfrac{2ab}{2 \cdot 3 \cdot a \cdot a \cdot b} = \dfrac{1}{3a}$

5. $\dfrac{12cd^2}{8c^2d} = \dfrac{4 \cdot 3 \cdot c \cdot d \cdot d}{4 \cdot 2 \cdot c \cdot c \cdot d} = \dfrac{3d}{2c}$

6. $\dfrac{(x - 2)\,(x + 3)}{(x + 3)} = x - 2$

7. $\dfrac{a - 6}{a^2 - 3a - 18} = \dfrac{(a - 6)}{(a - 6)\,(a + 3)}$

   $= \dfrac{1}{a + 3}$

8. $\dfrac{b + bc}{b} = \dfrac{b(1 + c)}{b} = 1 + c$

9. $\dfrac{a^2b}{c} \cdot \dfrac{c^2}{ab^2} = \dfrac{a \cdot a \cdot b \cdot c \cdot c}{c \cdot a \cdot b \cdot b} = \dfrac{ac}{b}$

10. $\dfrac{ab^2}{c^2} \div \dfrac{ac}{b} = \dfrac{ab^2}{c^2} \cdot \dfrac{b}{ac} = \dfrac{b^3}{c^3}$

11. $\dfrac{x^2 - 16}{x^2 + 6x + 8} \cdot \dfrac{x + 2}{x - 2}$

    $= \dfrac{(x + 4)\,(x - 4)}{(x + 4)\,(x + 2)} \cdot \dfrac{(x + 2)}{(x - 2)}$

    $= \dfrac{x - 4}{x - 2}$

12. $\dfrac{x^2 - 16}{x^2 + 6x + 8} \div \dfrac{4 - x}{x + 2}$

    $= \dfrac{(x + 4)\,(x - 4)}{(x + 4)\,(x + 2)} \cdot \dfrac{x + 2}{-1(x - 4)}$

    $= -1$

13. a) $\dfrac{2\frac{2}{3}\text{ yd}}{5 \text{ ft}} = \dfrac{\frac{8}{3}\text{ yd}}{5 \text{ ft}} = \dfrac{\frac{8}{3}\text{ yd} \cdot \frac{3 \text{ ft}}{1 \text{ yd}}}{5 \text{ ft}}$

    $= \dfrac{8 \text{ ft}}{5 \text{ ft}} = \dfrac{8}{5}$

    b) $\dfrac{150 \text{ months}}{15 \text{ years}} = \dfrac{150 \text{ months}}{15 \text{ yr} \cdot \frac{12 \text{ mon}}{1 \text{ yr}}}$

    $= \dfrac{150 \text{ months}}{180 \text{ month}} = \dfrac{15}{18} = \dfrac{3 \cdot 5}{3 \cdot 6} = \dfrac{5}{6}$

14. $1{,}000{,}000 \text{ Tbsp} \cdot \dfrac{1 \text{ cup}}{16 \text{ Tbsp}}$

    $\cdot \dfrac{1 \text{ pint}}{2 \text{ cups}} \cdot \dfrac{1 \text{ qt}}{2 \text{ pints}} \cdot \dfrac{1 \text{ gal}}{4 \text{ qts}}$

    $= 3906.25 \text{ gallons}$

15. $5 \text{ ft. } 2 \text{ inches} = 62 \text{ inches}$
    $62 - 15 = 47 \text{ inches}$

    $\dfrac{47 \text{ inches}}{14 \text{ years}} \cdot \dfrac{1 \text{ ft}}{12 \text{ inches}}$

    $\cdot \dfrac{1 \text{ mile}}{5280 \text{ ft}} \cdot \dfrac{1 \text{ yr}}{365 \text{ days}} \cdot \dfrac{1 \text{ day}}{24 \text{ hours}}$

    $\approx 6.048 \times 10^{-9} \text{ miles per hour}$

16. $\dfrac{a}{15} = \dfrac{12}{35}$, $35a = 15 \cdot 12$,

    $35a = 180$

    $a \approx 5.143$

17. $\dfrac{x + 1}{8} = \dfrac{2x - 1}{14}$

    $14(x + 1) = 8(2x - 1)$

    $14x + 14 = 16x - 8$

    $2x = 22, \ x = 11$

18. $\dfrac{6}{5} = \dfrac{9.6}{n}$, $6n = 5 \cdot 9.6$,

    $6n = 48, \ n = 8$

    $\dfrac{6}{5} = \dfrac{p}{11}$, $66 = 5p, \ p = 13.2$

Mid Chapter 4 Test

19. $m = \dfrac{15 - (-6)}{10 - (-4)} = \dfrac{21}{14} = \dfrac{3}{2}$

    $Y = \dfrac{3}{2}x + b, \ 15 = \dfrac{3}{2} \cdot 10 + b,$

    $b = 0$

    $y = \dfrac{3}{2}x,$ proportional.

20. $m = \dfrac{\$1.50 - \$0.60}{3 - 1} = \dfrac{\$0.90}{2} = \$0.45$

    $\$1.50 = \$0.45(3) + b, \ b = \$0.15$

    $y = \$0.45x + \$0.15,$

    Not porportional.

21. a) $4\,\text{ft} \cdot 100\,\text{lb} = 400\,\text{lb.ft.}$

    b) $80\,\text{lb} \cdot x = 400\,\text{lbft} \cdot x = 5\,\text{ft.}$
       (from center point or
       fulcrum)

22. $y = \dfrac{100}{x};$ $\ y = $ Cookies per child

    $x = $ No. of cookies

    $k = 100$ cookies

23. $\$10 = k(13)^2, \ k = \dfrac{\$10}{169\ \text{in}^2}$

    $p = \dfrac{\$10}{169\ \text{in}^2} \cdot 16^2\ \text{in}^2 \approx \$15.15$

24. $1350 = \dfrac{k}{1^2}, \ \ k = 1350$

    $n = \dfrac{1350}{4^2} \approx 84$ nails

Section 4.4

1. a) $\dfrac{2}{5} + \dfrac{x}{5} = \dfrac{2 + x}{5}$

    b) $\dfrac{2}{3x} - \dfrac{5}{3x} = \dfrac{2 - 5}{3x} = \dfrac{-3}{3x} = \dfrac{-1}{x}$

    c) $\dfrac{4}{x + 1} + \dfrac{x^2}{x + 1} = \dfrac{4 + x^2}{x + 1}, \ x \neq -1$

    d) $\dfrac{2}{x^2 - 1} - \dfrac{x + 1}{x^2 - 1} = \dfrac{2 - (x + 1)}{x^2 - 1}$

    $= \dfrac{2 - x - 1}{x^2 - 1} = \dfrac{1 - x}{x^2 - 1} = \dfrac{-1(x - 1)}{(x + 1)(x - 1)}$

    $= -\dfrac{1}{x + 1}, \ x \neq -1, 1$

3. a) $18 = 3 \cdot 6, \ 24 = 4 \cdot 6$

    Common denominator $= 3 \cdot 4 \cdot 6 = 72$

    b) $y^2 = y \cdot y$

    Common denominator $= y^2$

    $y \neq 0$

    c) Common denominator $= a \cdot b$

    $a \neq 0, \ b \neq 0$

    d) $x^2 - 25 = (x - 5)(x + 5)$

    Common denominator $= (x - 5)(x + 5)$

    $x \neq -5, 5$

    e) $x^2 + 5x + 6 = (x + 2)(x + 3)$

    $x^2 + 2x = x(x + 2), \ x \neq -3, -2, 0$

    Common denominator $= x(x + 2)(x + 3)$

5. a) $\dfrac{5}{18} + \dfrac{7}{24} = \dfrac{5 \cdot 4}{18 \cdot 4} + \dfrac{7 \cdot 3}{24 \cdot 3}$

    $= \dfrac{20}{72} + \dfrac{21}{72} = \dfrac{41}{72}$

    b) $\dfrac{7}{y^2} - \dfrac{2}{y} = \dfrac{7}{y^2} - \dfrac{2 \cdot y}{y \cdot y}$

    $= \dfrac{7}{y^2} - \dfrac{2y}{y^2} = \dfrac{7 - 2y}{y^2}$

    $y \neq 0$

---

c) $\dfrac{3}{a} - \dfrac{2}{b} + \dfrac{7}{a} - \dfrac{3}{b} = \dfrac{3b}{ab} - \dfrac{2a}{ab} + \dfrac{7b}{ab} - \dfrac{3a}{ab}$

$= \dfrac{3b - 2a + 7b - 3a}{ab} = \dfrac{10b - 5a}{ab}$

$a \neq 0, \ b \neq 0$

d) $\dfrac{4}{x - 5} - \dfrac{3x}{x^2 - 25}$

$= \dfrac{4(x + 5)}{(x - 5)(x + 5)} - \dfrac{3x}{(x - 5)(x + 5)}$

$= \dfrac{4x + 20 - 3x}{(x - 5)(x + 5)} = \dfrac{x + 20}{(x - 5)(x + 5)}$

$x \neq -5, 5$

e) $\dfrac{x}{x^2 + 5x + 6} - \dfrac{2x}{x^2 + 2x}$

$= \dfrac{x \cdot x}{x(x + 2)(x + 3)} - \dfrac{2x(x + 3)}{x(x + 2)(x + 3)}$

$= \dfrac{x^2 - 2x(x + 3)}{x(x + 2)(x + 3)} = \dfrac{x^2 - 2x^2 - 6x}{x(x + 2)(x + 3)}$

$= \dfrac{-x^2 - 6x}{x(x + 2)(x + 3)} = \dfrac{x(-x - 6)}{x(x + 2)(x + 3)}$

$= \dfrac{-x - 6}{(x + 2)(x + 3)}, \ x \neq -3, -2, 0$

7. $\dfrac{5}{2b} + \dfrac{7}{6a} = \dfrac{5 \cdot 3a}{2b \cdot 3a} + \dfrac{7b}{6ab} = \dfrac{15a + 7b}{6ab}$

    $a \neq 0, \ b \neq 0$

9. $\dfrac{x}{x - 3} + \dfrac{1}{x} = \dfrac{x \cdot x}{x(x - 3)} + \dfrac{1 \cdot (x - 3)}{x(x - 3)}$

    $= \dfrac{x^2 + x - 3}{x(x - 3)}, \ x \neq 0, 3$

11. $\dfrac{x}{x + 1} - \dfrac{3}{x - 1}$

    $= \dfrac{x(x - 1)}{(x + 1)(x - 1)} - \dfrac{3(x + 1)}{(x + 1)(x - 1)}$

    $= \dfrac{x^2 - x - 3x - 3}{(x + 1)(x - 1)} = \dfrac{x^2 - 4x - 3}{(x + 1)(x - 1)}$

    $x \neq -1, 1$

13.
$$\frac{x}{x-3} + \frac{3}{x^2 - 6x + 9}$$

$$= \frac{x}{x-3} + \frac{3}{(x-3)(x-3)}$$

$$= \frac{x(x-3)}{(x-3)(x-3)} + \frac{3}{(x-3)(x-3)}$$

$$= \frac{x^2 - 3x + 3}{(x-3)(x-3)}, \; x \neq 3$$

15.
$$\frac{1}{1-x} + \frac{x^2}{x-1} = -\frac{1}{x-1} + \frac{x^2}{x-1}$$

$$= \frac{-1+x^2}{x-1} = \frac{x^2-1}{x-1} = \frac{(x+1)(x-1)}{x-1}$$

$$= x+1, \; x \neq 1$$

17.
$$\frac{2}{2a+ab} - \frac{3}{2b+b^2} = \frac{2}{a(2+b)} - \frac{3}{b(2+b)}$$

$$= \frac{2b}{a(2+b)b} - \frac{3a}{b(2+b)a} = \frac{2b-3a}{ab(2+b)}$$

$$a \neq 0, \; b \neq -2, 0$$

19.
$$\frac{5}{x^2+x} + \frac{x}{x^2-2x-3}$$

$$= \frac{5}{x(x+1)} + \frac{x}{(x+1)(x-3)}$$

$$= \frac{5(x-3)}{x(x+1)(x-3)} + \frac{x \cdot x}{x(x+1)(x-3)}$$

$$= \frac{5x-15+x^2}{x(x+1)(x-3)} = \frac{x^2+5x-15}{x(x+1)(x-3)}$$

$$x \neq -1, 0, 3$$

21.
$$e^x \approx 1 + x + \frac{x^2}{2} + \frac{x^3}{6} + \frac{x^4}{24}$$

$$e^x \approx \frac{1 \cdot 24}{24} + \frac{x \cdot 24}{24} + \frac{x^2 \cdot 12}{2 \cdot 12} + \frac{x^3 \cdot 4}{6 \cdot 4} + \frac{x^4}{24}$$

$$e^x \approx \frac{24 + 24x + 12x^2 + 4x^3 + x^4}{24}$$

$$e^x \approx \frac{x^4 + 4x^3 + 12x^2 + 24x + 24}{24}$$

23.
$$P = \frac{RT}{v-b} - \frac{a}{v^2} = \frac{RTv^2}{(v-b)v^2} - \frac{a(v-b)}{v^2(v-b)}$$

$$P = \frac{RTv^2 - av + ab}{v^2(v-b)}$$

25.
$$\frac{1}{t} = \frac{1}{t_1} + \frac{1}{t_2} = \frac{t_2}{t_1 \cdot t_2} + \frac{t_1}{t_1 \cdot t_2}$$

$$\frac{1}{t} = \frac{t_2 + t_1}{t_1 t_2}$$

27.
$$\frac{1}{c} = \frac{1}{c_1} + \frac{1}{c_2} = \frac{c_2}{c_1 \cdot c_2} + \frac{c_1}{c_1 \cdot c_2}$$

$$\frac{1}{c} = \frac{c_2 + c_1}{c_1 c_2}$$

29.
$$\frac{1}{m} = \frac{1}{m_1} + \frac{1}{m_2} = \frac{m_2}{m_1 \cdot m_2} + \frac{m_1}{m_1 \cdot m_2}$$

$$\frac{1}{m} = \frac{m_2 + m_1}{m_1 m_2}$$

31. $t_1 = 5, \quad t_2 = 6$

$$\frac{1}{t} = \frac{t_2 + t_1}{t_1 t_2}, \; \frac{6+5}{30} = \frac{11}{30}$$

$$t = \frac{30}{11} \approx 2.727 \text{ hrs}$$

33. $367 = 45 \cdot t_1, \quad t_1 \approx 8.156 \text{ hrs}$

$367 = 60 \cdot t_2, \quad t_2 \approx 6.117 \text{ hrs}$

Total time $= t_1 + t_2 \approx 14.272 \text{ hrs}$

$2(367) = r \cdot 14.272, \; r \approx 51.429 \text{ mph}$

35. $v_1 = 55, \quad v_2 = 65$

$$v_{avg} = \frac{2v_1 v_2}{v_1 + v_2} \text{ (from ex. 10)}$$

$$v_{avg} = \frac{2 \cdot 55 \cdot 65}{55 + 65} = \frac{7150}{120} \approx 59.583 \text{ mph}$$

37. $m_1 = 5 \text{ min}, \quad m_2 = 9 \text{ min}$

$$m = \frac{m_1 m_2}{m_1 + m_2}$$

$$m = \frac{5 \cdot 9}{5 + 9} = \frac{45}{14} \approx 3.214 \text{ min}$$

39. $t\left(\dfrac{1}{t_1} + \dfrac{1}{t_2} + \dfrac{1}{t_3}\right)$

$= t\left(\dfrac{t_2 t_3}{t_1 t_2 t_3} + \dfrac{t_1 t_3}{t_1 t_2 t_3} + \dfrac{t_1 t_2}{t_1 t_2 t_3}\right)$

$t\left(\dfrac{t_2 t_3 + t_1 t_3 + t_1 t_2}{t_1 t_2 t_3}\right) = 1$

$t = \dfrac{t_1 t_2 t_3}{t_2 t_3 + t_1 t_3 + t_1 t_2}$

This is not the sum of $t_1 + t_2 + t_3$.

41. $\dfrac{7 + 1}{\dfrac{1}{7} + 1} = \dfrac{8}{\dfrac{1}{7} + \dfrac{7}{7}} = \dfrac{8}{\dfrac{8}{7}} = 8 \div \dfrac{8}{7}$

$= \dfrac{8}{1} \cdot \dfrac{7}{8} = 7$

43. $a = \dfrac{V}{\dfrac{4}{3}\pi b^2} = \dfrac{V}{\dfrac{4\pi b^2}{3}} = \dfrac{V}{1} \cdot \dfrac{3}{4\pi b^2}$

$a = \dfrac{3V}{4\pi b^2}$

45. $\dfrac{x - \dfrac{x}{2}}{2 + \dfrac{x}{3}} = \dfrac{\dfrac{2x}{2} - \dfrac{x}{2}}{\dfrac{2 \cdot 3}{3} + \dfrac{x}{3}} = \dfrac{\dfrac{2x - x}{2}}{\dfrac{6 + x}{3}}$

$= \dfrac{x}{2} \cdot \dfrac{3}{6 + x} = \dfrac{3x}{2(x + 6)}$

47. $I = \dfrac{E}{R + \dfrac{r}{2}} = \dfrac{E}{\dfrac{2R}{2} + \dfrac{r}{2}} = \dfrac{E}{\dfrac{2R + r}{2}}$

$I = \dfrac{E}{1} \cdot \dfrac{2}{2R + r} = \dfrac{2E}{2R + r}$

49. a) $m = \dfrac{\dfrac{d}{2} - 0}{\dfrac{d}{2} - 0} = \dfrac{\dfrac{d}{2}}{\dfrac{d}{2}} = 1$

b) $m = \dfrac{\dfrac{d}{2} - 0}{\dfrac{d}{2} - d} = \dfrac{\dfrac{d}{2}}{\dfrac{d}{2} - \dfrac{2d}{2}}$

$= \dfrac{\dfrac{d}{2}}{-\dfrac{d}{2}} = -1$

c) Slopes are negative inverses. Angle is 90 degrees.

51. $\dfrac{\dfrac{1}{x + h} - \dfrac{1}{x}}{h} = \dfrac{\dfrac{x}{x(x + h)} - \dfrac{1(x + h)}{x(x + h)}}{h}$

$= \dfrac{\dfrac{x - x - h}{x(x + h)}}{h} = \dfrac{-h}{x(x + h)} \cdot \dfrac{1}{h} = \dfrac{-1}{x(x + h)}$

53. $x^2(-1)(x + 1)^{-2} + \dfrac{2x}{x + 1}$

$= \dfrac{-x^2}{(x + 1)^2} + \dfrac{2x}{x + 1}$

$= \dfrac{-x^2}{(x + 1)^2} + \dfrac{2x(x + 1)}{(x + 1)^2}$

$= \dfrac{-x^2 + 2x^2 + 2x}{(x + 1)^2} = \dfrac{x^2 + 2x}{(x + 1)^2}$

$= \dfrac{x(x + 2)}{(x + 1)^2}$

55. a) $\dfrac{a}{b} \boxdot \dfrac{1}{a} = \dfrac{1}{b}$, Multiply

b) $\dfrac{a}{b} \boxplus \dfrac{1}{a} = \dfrac{a}{b} \cdot \dfrac{a}{1} = \dfrac{a^2}{b}$, Divide

57. a)

$\dfrac{1}{a} \boxplus \dfrac{1}{b} = \dfrac{b}{ab} + \dfrac{a}{ab} = \dfrac{a + b}{ab}$, Add

b) $\dfrac{1}{a} \boxdot \dfrac{1}{b} = \dfrac{1}{ab}$, Multiply

Section 4.4

59. $\dfrac{a}{b} + \dfrac{c}{d} = \dfrac{ad}{bd} + \dfrac{bc}{bd} = \dfrac{ad + bc}{bd}$

$\dfrac{b}{a} + \dfrac{d}{c} = \dfrac{bc}{ac} + \dfrac{ad}{ac} = \dfrac{bc + ad}{ac}$

$\dfrac{ad + bc}{bd} \neq \dfrac{ad + bc}{ac}$

61. $\left(\dfrac{a}{b} + \dfrac{c}{d}\right)^{-1} = \left(\dfrac{ad + bc}{bd}\right)^{-1} = \dfrac{bd}{ad + bc}$

but $\left(\dfrac{a}{b}\right)^{-1} + \left(\dfrac{c}{d}\right)^{-1} = \dfrac{b}{a} + \dfrac{d}{c} = \dfrac{bc + ad}{ac}$

Section 4.5

1. $\dfrac{x^2 - 3x - 4}{x - 4}$ Defined for all
$\mathbb{R}$ except $x = 4$

$$\frac{x^2 - 3x - 4}{x - 4} = \frac{(x - 4)(x + 1)}{(x - 4)} = x + 1$$

3. $\dfrac{x^2 - 3x - 4}{x - 3}$ Defined for all
$\mathbb{R}$ except $x = 3$

$$x - 3 \overline{)x^2 - 3x - 4} = x - \frac{4}{x - 3}$$
$$\underline{-(x^2 - 3x)}$$
$$0x - 4$$

with quotient x above.

5. $\dfrac{x^2 - 3x - 4}{x - 2}$ Defined for all
$\mathbb{R}$ except $x = 2$

$$x - 2 \overline{)x^2 - 3x - 4} = x - 1 - \frac{6}{x - 2}$$
$$\underline{-(x^2 - 2x)}$$
$$-x - 4$$
$$\underline{-(-x + 2)}$$
$$-6$$

with quotient $x - 1$ above.

7. $\dfrac{x^2 - 3x - 4}{x - 1}$ Defined for all
$\mathbb{R}$ except $x = 1$

$$x - 1 \overline{)x^2 - 3x - 4} = x - 2 - \frac{6}{x - 1}$$
$$\underline{-(x^2 - x)}$$
$$-2x - 4$$
$$\underline{-(-2x + 2)}$$
$$-6$$

with quotient $x - 2$ above.

9. $\dfrac{x^2 - 3x - 4}{x}$ Defined for all
$\mathbb{R}$ except $x = 0$

$$\frac{x^2 - 3x - 4}{x} = \frac{x^2}{x} - \frac{3x}{x} - \frac{4}{x}$$
$$= x - 3 - \frac{4}{x}$$

11. $\dfrac{x^2 - 3x - 4}{x + 1}$ Defined for all
$\mathbb{R}$ except $x = -1$

$$\frac{x^2 - 3x - 4}{x + 1} = \frac{(x - 4)(x + 1)}{x + 1} = x - 4$$

13. **1.**

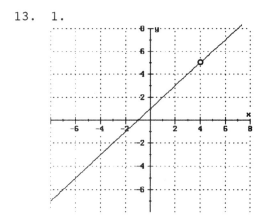

3.

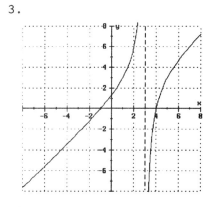

15. $x^2 - 3x - 4 = (x + 1)(x - 4)$
Quotient had a remainder of zero.

17.

x	$f(x) = x^2 - 3x - 4$
−1	$(-1)^2 - 3(-1) - 4 = 0$
0	$0^2 - 3 \cdot 0 - 4 = -4$
1	$1^2 - 3(1) - 4 = -6$
2	$2^2 - 3 \cdot 2 - 4 = -6$
3	$3^2 - 3 \cdot 3 - 4 = -4$
4	$4^2 - 3 \cdot 4 - 4 = 0$

Function values are the same as remainders.

Section 4.5

19. $f(-2) = (-2)^2 - 3(-2) - 4 = 6$

$f(5) = 5^2 - 3 \cdot 5 - 4 = 6$

$$\begin{array}{r} x - 5 \\ x + 2 \overline{\smash{)}x^2 - 3x - 4} \\ \underline{-(x^2 + 2x)} \\ -5x - 4 \\ \underline{-(-5x - 10)} \\ 6 \end{array} \qquad \begin{array}{r} x + 2 \\ x - 5 \overline{\smash{)}x^2 - 3x - 4} \\ \underline{-(x^2 - 5x)} \\ 2x - 4 \\ \underline{-(2x - 10)} \\ 6 \end{array}$$

Remainders equal 6.

21.

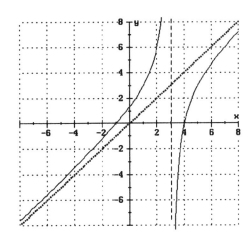

Graphs become nearly identical in the indicated regions.

23.

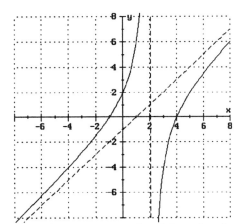

Graphs become nearly identical.

25. Inputs for which expressions are undefined exist after the division is done.

27. True.

29.
$$\begin{array}{r} x^2 + 2x + 4 \\ x - 2 \overline{\smash{)}x^3 + 0x^2 + 0x - 1} \\ \underline{-(x^3 - 2x^2)} \\ 2x^2 + 0x \\ \underline{-(2x^2 - 4x)} \\ 4x - 1 \\ \underline{-(4x - 8)} \\ 7 \end{array}$$

$$\frac{x^3 - 1}{x - 2} = x^2 + 2x + 4 + \frac{7}{x - 2}$$

31.
$$\begin{array}{r} x^2 + x + 1 \\ x - 1 \overline{\smash{)}x^3 + 0x^2 + 0x - 1} \\ \underline{-(x^3 - x^2)} \\ x^2 + 0x \\ \underline{-(x^2 - x)} \\ x - 1 \\ \underline{-(x - 1)} \\ 0 \end{array}$$

$$\frac{x^3 - 1}{x - 1} = \frac{(x - 1)(x^2 + x + 1)}{(x - 1)}$$

$$= x^2 + x + 1$$

Denominator is a factor of numerator.

33. $\dfrac{x^3 - 1}{x} = \dfrac{x^3}{x} - \dfrac{1}{x} = x^2 - \dfrac{1}{x}$

35.
$$\begin{array}{r} x^2 - x + 1 \\ x + 1 \overline{\smash{)}x^3 + 0x^2 + 0x - 1} \\ \underline{-(x^3 + x^2)} \\ -x^2 + 0x \\ \underline{-(-x^2 - x)} \\ x - 1 \\ \underline{-(x + 1)} \\ -2 \end{array}$$

$$\frac{x^3 - 1}{x + 1} = x^2 - x + 1 - \frac{2}{x + 1}$$

Section 4.5

37.

x	$f(x) = x^3 - 1$
−1	$(-1)^3 - 1 = -2$
0	$0^3 - 1 = -1$
1	$1^3 - 1 = 0$
2	$2^3 - 1 = 7$

Values correspond to remainders.

39. Remainder

41.

x	$f(x) = \dfrac{(x^3 + 3x^2 + 3x + 1)}{x + 1}$
−2	1
−1.8	0.64
−1.6	0.36
−1.4	0.16
−1.2	0.04
−1.0	undefined
−0.8	0.04
−0.6	0.16
−0.4	0.36
−0.2	0.64
0	1

43. They are the same.

45.

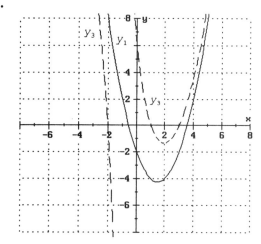

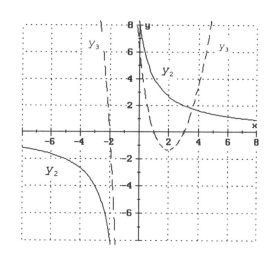

a) Part in $x > -1$ which resembles a parabola.

b) Nearly vertical section $x < -1$.

47.

$$
\begin{array}{r}
x^2 \;-\; xy \;+\; y^2 \\
x + y \overline{\smash{\big)}\, x^3 + 0x^2y + 0xy^2 + y^3} \\
\underline{-(x^3 + x^2y)} \\
-\,x^2y + 0xy^2 \\
\underline{-(-x^2y - xy^2)} \\
xy^2 + y^3 \\
\underline{-(xy^2 + y^3)} \\
0
\end{array}
$$

49.

$$
\begin{array}{r}
x^2 \;-\; x \;+\; 1 \\
x + y \overline{\smash{\big)}\, x^3 + 0x^2 + 0x + 1} \\
\underline{-(x^3 + x^2)} \\
-\,x^2 + 0x \\
\underline{-(-x^2 - x)} \\
x + 1 \\
\underline{-(x + 1)} \\
0
\end{array}
$$

or $\dfrac{(x + 1)(x^2 - x + 1)}{(x - 1)} = x^2 - x + 1$

Denominator is a factor.

51.

$$
x + 1 \overline{\smash{\big)}\, x^4 + 0x^3 + 0x^2 + 0x + 1}
$$

quotient: $x^3 - x^2 + x - 1 + \dfrac{2}{x+1}$

$$
\begin{array}{r}
x^3 - x^2 + x - 1 + \dfrac{2}{x+1} \\
x + 1 \overline{\smash{\big)}\, x^4 + 0x^3 + 0x^2 + 0x + 1} \\
\underline{-(x^4 + x^3)} \\
-x^3 + 0x^2 \\
\underline{-(-x^3 - x^2)} \\
x^2 + 0x \\
\underline{-(x^2 + x)} \\
-x + 1 \\
\underline{-(-x - 1)} \\
+2
\end{array}
$$

53.

$$
\begin{array}{r}
x^3 - 3x^2 + 3x - 1 \\
x - 1 \overline{\smash{\big)}\, x^4 - 4x^3 + 6x^2 - 4x + 1} \\
\underline{-(x^4 - x^3)} \\
-3x^3 + 6x^2 \\
\underline{-(-3x^3 + 3x^2)} \\
3x^2 - 4x \\
\underline{-(3x^2 - 3x)} \\
-x + 1 \\
\underline{-(-x + 1)} \\
0
\end{array}
$$

Denominator is a factor.

55.

$$
\begin{array}{r}
3x^2 - 2x + 2 - \dfrac{3}{x+1} \\
x + 1 \overline{\smash{\big)}\, 3x^3 + x^2 + 0x - 1} \\
\underline{-(3x^3 + 3x^2)} \\
-2x^2 + 0x \\
\underline{-(-2x^2 - 2x)} \\
2x - 1 \\
\underline{-(2x + 2)} \\
-3
\end{array}
$$

1. $\dfrac{3}{4}x + 5 = 23$

$\dfrac{3}{4}x = 23 - 5 = 18$

$\dfrac{4}{3} \cdot \dfrac{3}{4}x = 18 \cdot \dfrac{4}{3}$

$x = 24$

3. $L = \dfrac{\pi r \theta}{180}$

$\dfrac{180}{\pi\theta}L = \dfrac{\pi r \theta}{180} \cdot \dfrac{180}{\pi\theta}$

$\dfrac{180L}{\pi\theta} = r$

5. $x\left(\dfrac{1}{10} + \dfrac{1}{12}\right) = 1$

$x\left(\dfrac{6}{60} + \dfrac{5}{60}\right) = 1$

$x\left(\dfrac{11}{60}\right) = 1$

$\dfrac{60}{11} \cdot x \cdot \dfrac{11}{60} = 1 \cdot \dfrac{60}{11}$

$x = \dfrac{60}{11}$

7. $\dfrac{2n + 4}{7} = \dfrac{3n - 7}{4}$

$4(2n + 4) = 7(3n - 7)$

$8n + 16 = 21n - 49$

$65 = 13n$

$\dfrac{65}{13} = n$

$n = 5$

9. No variables were in the denominator.

11. $\dfrac{7}{x + 3} \overset{?}{=} \dfrac{x - 7}{8}$

$\dfrac{7}{(-7) + 3} \overset{?}{=} \dfrac{(-7) - 7}{8}$

$\dfrac{7}{-4} = \dfrac{-14}{8} = \dfrac{-7}{4}$

13. $\dfrac{3}{4} + \dfrac{1}{5} = \dfrac{1}{x}, \quad \dfrac{3}{4} + \dfrac{1}{5} \overset{?}{=} \dfrac{1}{\frac{20}{19}}$

$\dfrac{3}{4} + \dfrac{1}{5} \overset{?}{=} \dfrac{19}{20}$

$\dfrac{15}{20} + \dfrac{4}{20} = \dfrac{19}{20}$

15. $\dfrac{2}{X + 1} + \dfrac{3}{x} = -2$

$\dfrac{2}{(-3) + 1} + \dfrac{3}{(-3)} \overset{?}{=} -2$

$\dfrac{2}{-2} + \dfrac{3}{-3} = -2$

$-1 + -1 = -2$

17.

$(x - 1)(x + 2)$	$\dfrac{1}{x - 1}$	$\dfrac{1}{x + 2}$
	$\dfrac{(x - 1)(x + 2)}{x - 1}$	$\dfrac{(x - 1)(x + 2)}{x + 2}$

$= (x + 2) + (x - 1) = 2x + 1$

19. $2x^2\left(\dfrac{1}{2x^2} + \dfrac{3}{x}\right) = \dfrac{2x^2}{1} \cdot \dfrac{1 + 6x}{2x^2} = 1 + 6x$

21. $x(x + 1)\left(\dfrac{1}{x + 1} - \dfrac{2}{x}\right)$

$= \dfrac{x(x + 1)}{(x + 1)} - \dfrac{2x(x + 1)}{x}$

$= x - 2(x + 1)$

$= x - 2x - 2$

$= -x - 2$

23. $\dfrac{x}{4} + \dfrac{x}{6} = 15$

$24\left(\dfrac{x}{4} + \dfrac{x}{6}\right) = 15 \cdot 24$

$6x + 4x = 360$

$10x = 360$

$x = 36$

25. $\dfrac{3}{5} + \dfrac{2}{3} = \dfrac{1}{x}, \ x \neq 0$

$15x\left(\dfrac{3}{5} + \dfrac{2}{3}\right) = \dfrac{1}{x} \cdot 15x$

$9x + 10x = 15$

$19x = 15$

$x = \dfrac{15}{19}$

27. $\dfrac{1}{3} + \dfrac{1}{x} = \dfrac{1}{2}, \ x \neq 0$

$6x\left(\dfrac{1}{3} + \dfrac{1}{x}\right) = \dfrac{1}{2} \cdot 6x$

$2x + 6 = 3x$

$x = 6$

29. $\dfrac{5}{x} + \dfrac{2}{x} = \dfrac{4}{x}, \ x \neq 0$

$x\left(\dfrac{5}{x} + \dfrac{2}{x}\right) = \dfrac{4}{x} \cdot x$

$5 + 2 = 4$

$7 \neq 4$

False; no solution

$\{\ \}$

31. $\dfrac{1}{2} - \dfrac{1}{x} = \dfrac{1}{2x}, \ x \neq 0$

$2x\left(\dfrac{1}{2} - \dfrac{1}{x}\right) = \dfrac{1}{2x} \cdot 2x$

$x - 2 = 1$

$x = 3$

33. $\dfrac{4}{2x - 1} = \dfrac{1}{X - 1}, \ X \neq \dfrac{1}{2}, \ 1$

$4(X - 1) = 1(2X - 1)$

$4X - 4 = 2X - 1$

$2X = 3$

$X = \dfrac{3}{2}$

35. $\dfrac{18}{X^2} + \dfrac{9}{X} - 2 = 0, \ X \neq 0$

$X^2\left(\dfrac{18}{X^2} + \dfrac{9}{X} - 2\right) = 0 \cdot X^2$

$18 + 9X - 2X^2 = 0$

$-2X^2 + 9X + 18 = 0$

$-2X^2 + 12X - 3X + 18 = 0$

	x	-6
$-2x$	$-2x^2$	$+12x$
-3	$-3x$	$+18$

$(x - 6)(-2x - 3) = 0$

$x - 6 = 0$

$x = 6$

$-2x - 3 = 0$

$x = -\dfrac{3}{2}$

$\left\{-\dfrac{3}{2}, \ 6\right\}$

37. $\dfrac{1}{x - 2} - 4 = \dfrac{3 - x}{x - 2}, \quad x \neq 2$

$(x - 2)\left(\dfrac{1}{x - 2} - 4\right) = \dfrac{3 - x}{x - 2} \cdot x - 2$

$1 - 4(x - 2) = 3 - x$

$1 - 4x + 8 = 3 - x$

$6 = 3x$

$2 = x$ but $x \neq 2$, so $\{\}$

39. $\dfrac{3}{x-1} + \dfrac{5}{2x+2} = \dfrac{3}{2}, \; x \neq -1, 1$

$(x-1)(2x+2)\left(\dfrac{3}{x-1} + \dfrac{5}{2x+2}\right)$

$= (x-1)(2x+2)\dfrac{3}{2}$

$6x + 6 + 5x - 5 = 3(x-1)(x+1)$

$11x + 1 = 3x^2 - 3$

$3x^2 - 12X + X - 4 = 0$

	x	**-4**
3x	$3x^2$	$-12x$
1	x	-4

$(x-4)(3x+1) = 0$

$x - 4 = 0, \; x = 4$

$3x + 1 = 0, \; x = -\dfrac{1}{3}$

$\left\{-\dfrac{1}{3}, 4\right\}$

41. $\dfrac{x}{6-x} = \dfrac{2}{x-3}, \; x \neq 3, 6$

$x(x-3) = 2(6-x)$

$x^2 - 3x = 12 - 2x$

$x^2 - x - 12 = 0$

$(x-4)(x+3) = 0$

$x - 4 = 0, \; x = 4$

$x + 3 = 0, \; x = -3$

$\{-3, 4\}$

43. $\dfrac{x+1}{x-1} = \dfrac{1}{1-x}, \; x \neq 1$

$(x-1)(1-x)\left(\dfrac{x+1}{x-1}\right) = \dfrac{1}{1-x}(x-1)(1-x)$

$(1-x)(x+1) = x - 1$

$-1(x-1)(x+1) = x - 1$

$-1(x^2 - 1) = x - 1$

$-x^2 + 1 = x - 1$

$-x^2 - x + 2 = 0$

$x^2 + x - 2 = 0$

$(x+2)(x-1) = 0$

$x + 2 = 0, \; x = -2$

$x - 1 = 0, \; x = 1, \text{ but } x \neq 1$

$x = -2$

45. $15x(x+2)\left(\dfrac{1}{x+2} + \dfrac{2}{x}\right) = \left(\dfrac{11}{15}\right)15x(x+2)$

$15x + 15(x+2) \cdot 2 = 11x(x+2)$

$15x + 30x + 60 = 11x^2 + 22x$

$45x + 60 = 11x^2 + 22x$

$11x^2 - 23x - 60 = 0$

Did not distribute

$15x(x+2) \text{ over } \dfrac{2}{x}.$

47. $15x(x+2)\left(\dfrac{1}{x+2} + \dfrac{2}{x}\right) = \left(\dfrac{11}{15}\right)15x(x+2)$

$15x + 30(x+2) = 11x(x+2)$

$15x + 30x + 60 = 11x^2 + 22$

$11x^2 - 23x - 60 = 0$

Forgot to multiply by 2 in the second term.

49. $\dfrac{1}{6} + \dfrac{1}{x} = \dfrac{1}{2}$

$6x\left(\dfrac{1}{6} + \dfrac{1}{x}\right) = \dfrac{1}{2} \cdot 6x$

$x + 6 = 3x$

$6 = 2x$

$x = 3 \text{ minutes}$

51. $\dfrac{1}{20} + \dfrac{1}{x} = \dfrac{1}{12}$

$60x\left(\dfrac{1}{20} + \dfrac{1}{x}\right) = \dfrac{1}{12} \cdot 60x$

$3x + 60 = 5x$

$60 = 2x$

$x = 30 \text{ minutes}$

53. $\dfrac{1}{10,000} + \dfrac{1}{4,000} = \dfrac{1}{R}$

$20,000R\left(\dfrac{1}{10,000} + \dfrac{1}{4,000}\right) = \dfrac{1}{R} \cdot 20,000R$

$2R + 5R = 20,000$

$7R = 20,000$

$R = \dfrac{20,000}{7} \approx 2857 \text{ ohms}$

Section 4.6

55. $\dfrac{1}{5} + \dfrac{1}{3} + \dfrac{1}{3} = \dfrac{1}{x}$

$15x\left(\dfrac{1}{5} + \dfrac{1}{3} + \dfrac{1}{3}\right) = \dfrac{1}{x} \cdot 15x$

$3x + 5x + 5x = 15$

$13x = 15$

$x = \dfrac{15}{13} \approx 1.154 \text{ minutes}$

57. $2\ell + 2\omega = 39.9, \quad \dfrac{\ell}{\omega} = \dfrac{5}{2}, \quad \ell = \dfrac{5}{2}\omega$

$2\left(\dfrac{5}{2}\omega\right) + 2\omega = 39.9$

$5\omega + 2\omega = 39.9$

$7\omega = 39.9$

$\omega = 5.7 \text{ cm}$

$\ell = \dfrac{5}{2} \cdot 5.7 \approx 14.3 \text{ cm}$

59. $\dfrac{\omega}{\ell} = \dfrac{3}{7}, \quad \omega = \dfrac{3}{7}\ell, \ 2\ell + 2\omega = 44.4$

$2\ell + 2\left(\dfrac{3}{7}\ell\right) = 44.4$

$2\ell + \dfrac{6}{7}\ell = 44.4, \quad \dfrac{14}{7}\ell + \dfrac{6}{7}\ell = 44.4$

$\dfrac{20}{7}\ell = 44.4, \quad \ell = 44.4 \cdot \dfrac{7}{20} \approx 15.5 \text{ cm}$

$\omega = \dfrac{3}{7}(15.5) \approx 6.6 \text{ cm}$

61. $\dfrac{1}{a} + \dfrac{1}{b} = \dfrac{1}{c}$

$abc\left(\dfrac{1}{a} + \dfrac{1}{b}\right) = \dfrac{1}{c}abc$

$bc + ac = ab$

$bc = ab - ac$

$bc = a(b - c)$

$a = \dfrac{bc}{b - c}$

63. $\dfrac{1}{a} + \dfrac{1}{b} = \dfrac{1}{c}$

$abc\left(\dfrac{1}{a} + \dfrac{1}{b}\right) = \dfrac{1}{c}abc$

$c(b + a) = ab$

$c = \dfrac{ab}{a + b}$

65. $I = \dfrac{E}{R + r}$

$I(R + r) = E$

$R + r = \dfrac{E}{I}$

$R = \dfrac{E}{I} - r$

67. $\left\{\dfrac{1}{2}, 2\right\}$

1. a) $\dfrac{27xy^2}{15x^2y} = \dfrac{3 \cdot a \cdot \overset{\checkmark}{x} \cdot \overset{\checkmark}{y} \cdot \overset{\checkmark}{y}}{3 \cdot 5 \cdot \underset{\checkmark}{x} \cdot \underset{\checkmark}{x} \cdot \underset{\checkmark}{y}} = \dfrac{9y}{5x}$

   b) $\dfrac{ac}{a(b + c)} = \dfrac{c}{b + c}$

   c) $\dfrac{a + b}{a^2 + 2ab + b^2} = \dfrac{a + b}{(a + b)(a + b)}$

   $= \dfrac{1}{a + b}$

3. a) $\dfrac{2 - x}{x - 2} = \dfrac{-1(x - 2)}{(x - 2)} = -1$

   b) $\dfrac{3(1 + \sqrt{3})}{3} = 1 + \sqrt{3}$

   c) $\dfrac{6 - 2x}{6} = \dfrac{2(3 - x)}{2 \cdot 3} = \dfrac{3 - x}{3}$

5. a) $\dfrac{3x^2 - 12}{x + 2} = \dfrac{3(x^2 - 4)}{x + 2}$

   $= \dfrac{3(x + 2)(x - 2)}{x + 2}$

   $= 3(x - 2)$

   b) $\dfrac{a - b}{(a + b)(a - b)} = \dfrac{1}{a + b}$

   c) $8x\left(\dfrac{1}{4x} + \dfrac{1}{2}\right) = \dfrac{8x}{4x} + \dfrac{8x}{2} = 2 + 4x$

7. a) $\dfrac{x^2 - 5x - 6}{x^2 - 4x - 5} = \dfrac{(x - 6)(x + 1)}{(x - 5)(x + 1)}$

   $= \dfrac{x - 6}{x - 5}$

   b) $\dfrac{4x^2 - 1}{2x^2 + 5x + 2} = \dfrac{(2x - 1)(2x + 1)}{(2x + 1)(x + 2)}$

   $= \dfrac{2x - 1}{x + 2}$

   c) $\dfrac{x^2 + 3x - 4}{x^2 - 16} = \dfrac{(x + 4)(x - 1)}{(x + 4)(x - 4)}$

   $= \dfrac{x - 1}{x - 4}$

9. $\dfrac{x + y}{y} = \dfrac{x}{y} + \dfrac{y}{y} = \dfrac{x}{y} + 1$, c

11. $\dfrac{x}{x + y} = \dfrac{x}{y + x}$, a

13. $\dfrac{a}{a - b} = \dfrac{a + b - b}{a - b} = \dfrac{a - b + b}{a - b}$

   $= \dfrac{a - b}{a - b} + \dfrac{b}{a - b} = 1 + \dfrac{b}{a - b}$, b

15. $(x - 1) = 0$ or $(x + 3) = 0$

   $x = 1$ or $\qquad x = -3$

17. $\dfrac{4y^2}{9x^2} \cdot \dfrac{3x}{8y} = \dfrac{4 \cdot 3 \cdot x \cdot y^2}{8 \cdot 9 \cdot x^2 \cdot y} = \dfrac{y}{6x}$

19. $15x\left(\dfrac{1}{3x} + \dfrac{2}{5x}\right) = \dfrac{15x}{3x} + \dfrac{30x}{5x}$

   $= 5 + 6 = 11$

21. $\dfrac{x^2 + 5x + 4}{x^2 - 16} \cdot \dfrac{2x - 8}{1 - x^2}$

   $= \dfrac{(x + 1)(x + 4)}{(x - 4)(x + 4)} \cdot \dfrac{2(x - 4)}{(1 - x)(1 + x)}$

   $= \dfrac{2(x + 1)}{(1 - x)(x + 1)} = \dfrac{2}{1 - x}$

23. $\dfrac{1}{x - 3} + \dfrac{x}{x - 3} = \dfrac{1 + x}{x - 3} = \dfrac{x + 1}{x - 3}$

25. $\dfrac{x - 2}{x - 1} - \dfrac{x}{x + 2}$

   $= \dfrac{(x - 2)(x + 2)}{(x - 1)(x + 2)} - \dfrac{x(x - 1)}{(x - 1)(x + 2)}$

   $= \dfrac{x^2 - 4 - x^2 + x}{(x - 1)(x + 2)} = \dfrac{x - 4}{(x - 1)(x + 2)}$

27. $\dfrac{3 \text{ qts}}{4 \text{ gal}} \cdot \dfrac{1 \text{ gal}}{4 \text{ qts}} = \dfrac{3}{16}$, 3 to 16

29. $\dfrac{1728 \text{ in}^3}{144 \text{ in}^2} = \dfrac{1728}{144} \cdot \text{in}^{3-2} = 12 \text{ in.}$

31. $\dfrac{2500 \text{ miles}}{12.5 \frac{\text{miles}}{\text{gal}}} = \dfrac{2500 \text{ miles}}{12.5} \cdot \dfrac{\text{gal}}{\text{miles}}$

    $= 200 \text{ gal.}$

33. y approaches $-\dfrac{3}{4}$

35. Graph becomes nearly vertical. y approaches $-\infty, +\infty$.

37. $\dfrac{6 - x}{\Box} = -1$, $6 - x = -1 \cdot \Box$,

    $\dfrac{6 - x}{-1} = \Box$, $\Box = -1(6 - x) = x - 6$

39. Per means division & division is multiplying by the reciprocal.

41. $1{,}000{,}000 \text{ sec} \cdot \dfrac{1 \text{ hr}}{3600 \text{ sec}} \cdot \dfrac{1 \text{ day}}{24 \text{ hr}}$

    $\approx 11.6 \text{ days} \approx 10 \text{ days}$

43. $1{,}000{,}000 \text{ hr} \cdot \dfrac{1 \text{ day}}{24 \text{ hr}} \cdot \dfrac{1 \text{ yr}}{365 \text{ day}}$

    $\approx 114 \text{ yr} \approx 100 \text{ yrs}$

45. $\dfrac{21 \text{ in}}{9 \text{ mon}} \cdot \dfrac{1 \text{ mon}}{30 \text{ day}} \cdot \dfrac{1 \text{ day}}{24 \text{ hr}} \cdot \dfrac{1 \text{ ft}}{12 \text{ in}} \cdot \dfrac{1 \text{ mile}}{5280 \text{ ft}}$

    $\approx 5.115 \times 10^{-8} \text{ MPH}$

47. $\dfrac{48 \text{ in}}{18 \text{ yrs}} \cdot \dfrac{1 \text{ yr}}{365 \text{ day}} \cdot \dfrac{1 \text{ day}}{24 \text{ hr}} \cdot \dfrac{1 \text{ ft}}{12 \text{ in}} \cdot \dfrac{1 \text{ mile}}{5280 \text{ ft}}$

    $\approx 4.805 \times 10^{-9} \text{ MPH}$

49. $\dfrac{21 \text{ in}}{x} = \dfrac{48 \text{ in}}{18 \text{ yr}}$, $x = \dfrac{21 \cdot 18}{48}$

    $\approx 7.9 \text{ yrs}$

51. $\dfrac{4}{5} = \dfrac{x}{6}$ $\qquad \dfrac{4}{5} = \dfrac{y}{7}$

    $\dfrac{6 \cdot 4}{5} = x$ $\qquad \dfrac{7 \cdot 4}{5} = y$

    $x = \dfrac{24}{5} = 4.8$ $\quad y = \dfrac{28}{5} = 5.6$

53. $\dfrac{7.5}{6} = \dfrac{r}{8}$, $6r = 60$, $r = 10$

    $s = \sqrt{6^2 + 8^2} = \sqrt{100} = 10$

    $\dfrac{7.5}{6} = \dfrac{t}{10}$, $6t = 75$, $t = 12.5$

55. $(12, \$9)$, $(24, \$12)$

    $m = \dfrac{\$12 - 9}{24 - 12} = \dfrac{3}{12} = \dfrac{1}{4} = \$0.25$

    $y = \$0.25x + b$,

    $\$9 = \$0.25(12) + b$,

    $b = 6 = \text{Cost of developing}$

    $y = \$0.25x + \$6$

57. $(40, \$74.80)$, $(65, \$121.55)$

    $m = \dfrac{121.55 - 74.80}{65 - 40} = \dfrac{\$46.75}{25}$

    $= \$1.87$

    $\$74.80 = \$1.87(40) + b$, $b = 0$

    $y = \$1.87x$ Proportional

59. $D = k\sqrt{h}$, $\quad D = \text{Distance}$
    $\qquad\qquad\quad h = \text{Height}$

61. $V = \ell \cdot w \cdot h$, $\quad V = \text{Volume}$,
    $\qquad\qquad\qquad\quad \ell = \text{Length}$,
    $\qquad\qquad\qquad\quad w = \text{Width}$,
    $\qquad\qquad\qquad\quad h = \text{Height}$

63. $d = ks^2$

65. $s = k\sqrt{df}$

67. $y = \dfrac{90}{x}$, $\quad y = \text{No. quarters}$
    $\qquad\qquad\quad k = 90 \text{ credits}$

69. $y = \dfrac{\$10{,}000}{x}$, $k = \$10{,}000$
    $\quad y = \text{output} \quad x = \text{input}$

71. a) $r = \dfrac{300}{(t - 3)}$, rate, r, is based on total hrs less 3 hrs down time.

b)

Total hrs	Rate mph
4	$300;\ \dfrac{300}{4 - 3} = \dfrac{300}{1} = 300$
5	$150;\ \dfrac{300}{5 - 3} = \dfrac{300}{2} = 150$
6	$100;\ \dfrac{300}{6 - 3} = \dfrac{300}{3} = 100$
7	$75;\ \dfrac{300}{7 - 3} = \dfrac{300}{4} = 75$
8	60

c) Replacing t with t-3 causes a shift of 3 units to the right.

73. $r = \dfrac{300}{3.5 - 3} = \dfrac{300}{0.5} = 600\ \text{mph}$

$r = \dfrac{300}{3.1 - 3} = \dfrac{300}{0.1} = 3000\ \text{mph}$

$r = \dfrac{300}{3.01 - 3} = \dfrac{300}{0.01} = 30{,}000\ \text{mph}$

$r = \dfrac{300}{3 - 3} = \dfrac{300}{0} = \text{not possible}$

75. $h = \dfrac{V}{\dfrac{1}{3}\pi r^2} = \dfrac{V}{\dfrac{\pi r^2}{3}} = \dfrac{V}{1} \cdot \dfrac{3}{\pi r^2}$

$h = \dfrac{3V}{\pi r^2}$

77. $\dfrac{x - \dfrac{4}{x}}{1 - \dfrac{2}{x}} = \dfrac{\dfrac{x^2 - 4}{x}}{\dfrac{x - 2}{x}} = \dfrac{x^2 - 4}{x} \cdot \dfrac{x}{x - 2}$

$= \dfrac{x^2 - 4}{x - 2} = \dfrac{(x - 2)(x + 2)}{(x - 2)} = x + 2$

79. $\dfrac{1}{5.5} + \dfrac{1}{7} \overset{?}{=} \dfrac{1}{3}$

$\dfrac{7}{5.5 \cdot 7} + \dfrac{5.5}{5.5 \cdot 7} \overset{?}{=} \dfrac{1}{3}$

$\dfrac{12.5}{38.5} \overset{?}{=} \dfrac{1}{3}$

$\approx 0.3247 \neq 0.3333$

Will not change air fast enough.

81. $\dfrac{x^3 - 3x^2 + 3x - 1}{x}$, $x \neq 0$

$\dfrac{x^3}{x} - \dfrac{3x^2}{x} + \dfrac{3x}{x} - \dfrac{1}{x}$

$= x^2 - 3x + 3 - \dfrac{1}{x}$

83. $\dfrac{x^3 - 3x^2 + 3x - 1}{x - 1}$, $x \neq 1$

$$
\begin{array}{r}
x^2 - 2x + 1 \\
x - 1 \overline{\smash{)}\ x^3 - 3x^2 + 3x - 1} \\
\underline{-(x^3 - x^2)} \\
-2x^2 + 3x) \\
\underline{-(-2x^2 + 2x)} \\
x - 1 \\
\underline{-(x - 1)} \\
0
\end{array}
$$

$x^2 - 2x + 1$

85. $\dfrac{x^3 - 3x^2 + 3x - 1}{x - 2}$, $x \neq 2$

$$
\begin{array}{r}
x^2 - x + 1 \\
x - 2 \overline{\smash{)}\ x^3 - 3x^2 + 3x - 1} \\
\underline{-(x^3 - 2x^2)} \\
-x^2 + 3x) \\
\underline{-(-x^2 + 2x)} \\
x - 1 \\
\underline{-(x - 2)} \\
1
\end{array}
$$

$x^2 - x + 1 + \dfrac{1}{x - 2}$

Chapter 4 Review

87. $\dfrac{x^3 - 3x^2 + 3x - 1}{x - 3}$, $x \neq 3$

$$x - 3 \overline{\smash{\big)}\, \begin{array}{r} x^2 + 0x + 3 \\ x^3 - 3x^2 + 3x - 1 \end{array}}$$

$$\begin{array}{r}
\underline{-(x^3 - 3x^2)} \\
0x^2 + 3x) \\
\underline{-(0x^2 + 0x)} \\
3x - 1 \\
\underline{-(3x - 9)} \\
8
\end{array}$$

$x^2 + 3 + \dfrac{8}{x - 3}$

89.

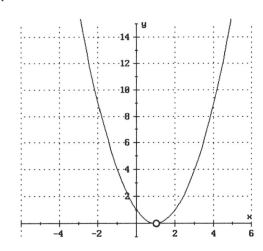

Denominator is a factor of the numerator.

91. $\dfrac{6}{x} = \dfrac{15}{32}$, $x \neq 0$

$15x = 192$

$\dfrac{192}{15} = x$

$x = 12.8$

93. $\dfrac{x}{2} = \dfrac{7}{x + 5}$, $x \neq 5$

$x^2 + 5x = 14$

$x^2 + 5x - 14 = 0$

$(x + 7)(x - 2) = 0$

$x + 7 = 0$, $x = -7$

$x - 2 = 0$, $x = 2$

$\{-7, 2\}$

95. $\dfrac{x - 2}{4} = \dfrac{x + 1}{3} - 3$

$12\left(\dfrac{x - 2}{4}\right) = 12\left(\dfrac{x + 1}{3} - 3\right)$

$3x - 6 = 4x + 4 - 36$

$3x - 6 = 4x - 32$

$26 = x$

$x = 26$

97. $\dfrac{1}{x} + \dfrac{5}{x + 2} = \dfrac{13}{3x}$, $x \neq -2, 0$

$3x(x + 2)\left(\dfrac{1}{x} + \dfrac{5}{x + 2}\right) = \dfrac{13}{3x} \cdot 3x(x + 2)$

$3(x + 2) + 15x = 13(x + 2)$

$3x + 6 + 15x = 13x + 26$

$18x + 6 = 13x + 26$

$5x = 20$

$x = 4$

99. $\dfrac{1}{x-1} + \dfrac{2}{x} = \dfrac{1}{6}$, $x \neq 0, 1$

$6x(x-1)\left(\dfrac{1}{x-1} + \dfrac{2}{x}\right)$

$= \dfrac{7}{6}(6x)(x-1)$

$6x + 12(x-1) = 7x(x-1)$

$6x + 12x - 12 = 7x^2 - 7x$

$18x - 12 = 7x^2 - 7x$

$7x^2 - 25x + 12 = 0$

	x	**-3**
7x	$7x^2$	-21x
-4	-4x	+12

$(7x - 4)(x - 3) = 0$

$7x - 4 = 0$, $x = \dfrac{4}{7}$

$x - 3 = 0$, $x = 3$

$\left\{\dfrac{4}{7}, 3\right\}$

101. a) $m = \dfrac{8-4}{5-10} = \dfrac{4}{-5}$

$8 = -\dfrac{4}{5}(5) + b$, $b = 12$

$y = -\dfrac{4}{5}x + 12$

b) $y \cdot x = 8 \cdot 5 = 40$

$y = \dfrac{40}{x}$

c)

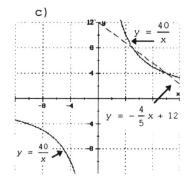

d) Both contain the points (5, 8) and (10, 4); a. is straight and b. is a hyperbola.

106

Chapter 4 Test

1. $\dfrac{ab^2c}{a^2bc^2} = \dfrac{b}{ac}$;
   $a \neq 0,\ b \neq 0,\ c \neq 0$

2. $\dfrac{b + 3}{b - 3}$, $b \neq 3$; Already simplified

3. $\dfrac{3ac}{15ac^2} = \dfrac{1}{5c}$; $a \neq 0,\ c \neq 0$

4. $\dfrac{15b^2c^3}{10b^3c} = \dfrac{3c^2}{2b}$; $b \neq 0,\ c \neq 0$

5. $\dfrac{x^2 - 25}{-x^2 + x + 20} = \dfrac{(x + 5)\,(x - 5)}{(x - 5)\,(-x - 4)}$;
   $x \neq -4,\ 5$
   $\dfrac{x + 5}{-x - 4} = -\dfrac{x + 5}{x + 4}$

6. $\dfrac{12 - 2a}{-a^2 + 3a + 18} = \dfrac{2(-a + 6)}{(-a + 6)\,(a + 3)}$
   $= \dfrac{2}{a + 3}$; $a \neq -3,\ 6$

7. $\dfrac{3x^2 - 7x + 2}{2x^2 - 5x + 2} = \dfrac{(3x - 1)\,(x - 2)}{(2x - 1)\,(x - 2)}$;
   $x \neq \dfrac{1}{2},\ 2$
   $= \dfrac{3x - 1}{2x - 1}$

8. $\dfrac{2x^2 + 7x + 6}{x^2 + 4x + 4} = \dfrac{(2x + 3)\,(x + 2)}{(x + 2)\,(x + 2)}$;
   $x \neq -2$
   $\dfrac{2x + 3}{x + 2}$

9. $\dfrac{12xy^2}{7y - y^2} \div \dfrac{6x^2}{7 - y} = \dfrac{12xy^2}{y(7 - y)} \cdot \dfrac{7 - y}{6x^2}$
   $= \dfrac{2y}{x}$; $x \neq 0,\ y \neq 0,\ 7$

10. $\dfrac{x^2 - 9}{x^2 + 4x + 3} \cdot \dfrac{x - 3}{x + 1}$
    $= \dfrac{(x + 3)\,(x - 3)}{(x + 3)\,(x + 1)} \cdot \dfrac{(x - 3)}{(x + 1)}$
    $= \dfrac{(x - 3)^2}{(x + 1)^2}$; $x \neq -3,\ -1$

11. $\dfrac{x^2 + 8x + 16}{x - 2} \cdot \dfrac{x^2 - 4}{x + 4}$
    $= \dfrac{(x + 4)\,(x + 4)}{(x - 2)} \cdot \dfrac{(x + 2)\,(x - 2)}{(x + 4)}$
    $= (x + 4)\,(x + 2)$ or $x^2 + 6x + 8$;
    $x \neq -4,\ 2$

12. $x(x + 1)\left(\dfrac{3}{x} + \dfrac{2}{x + 1}\right) = 3x + 3 + 2x$
    $= 5x + 3$; $x \neq -1,\ 0$

13. $\dfrac{2 - x}{x + 2} + \dfrac{x + 2}{x - 2}$
    $= \dfrac{(x - 2)\,(-1)\,(x - 2)}{(x + 2)\,(x - 2)} + \dfrac{(x + 2)\,(x + 2)}{(x + 2)\,(x - 2)}$
    $= \dfrac{(-1)\,(x^2 - 4x + 4) + x^2 + 4x + 4}{(x + 2)\,(x - 2)}$
    $= \dfrac{-x^2 + 4x - 4 + x^2 + 4x + 4}{(x + 2)\,(x - 2)}$
    $= \dfrac{8x}{(x + 2)\,(x - 2)}$; $x \neq -2,\ 2$

14. $\dfrac{x - 2}{x + 2} - \dfrac{x}{x(x + 2)}$
    $= \dfrac{x^2 - 2x}{x(x + 2)} - \dfrac{x}{x(x + 2)}$
    $= \dfrac{x^2 - 3x}{x(x + 2)} = \dfrac{x(x - 3)}{x(x + 2)}$
    $= \dfrac{x - 3}{x + 2}$; $x \neq -2,\ 0$

15. $\dfrac{10{,}000 \text{ cm}^3}{10 \text{ cm}} = 1000 \text{ cm}^2$

Chapter 4 Test

16.
$$\frac{30\,\frac{\text{miles}}{\text{gal}}}{60\,\frac{\text{miles}}{\text{hr}}} = \frac{1}{2} \cdot \frac{\text{miles}}{\text{gal}} \cdot \frac{\text{hr}}{\text{mile}}$$
$$= \frac{1}{2}\,\frac{\text{hr}}{\text{gal}}$$

17.
$$\frac{\dfrac{3}{x} - x}{x + \dfrac{2}{x}} = \frac{\dfrac{3 - x^2}{x}}{\dfrac{x^2 + 2}{x}}$$
$$= \frac{3 - x^2}{x} \cdot \frac{x}{x^2 + 2}$$
$$= \frac{3 - x^2}{x^2 + 2}\,;\ x \neq 0$$

18.
$$\frac{\dfrac{x^2 - 3x}{3}}{\dfrac{9 - x^2}{x}} = \frac{x(x - 3)}{3} \cdot \frac{x}{(-1)\,(x + 3)\,(x - 3)}$$
$$= \frac{x^2}{-3(x + 3)} = -\frac{x^2}{3(x + 3)}\,;$$
$$x \neq -3,\ 0,\ 3$$

19.
$$\frac{7\,\text{lbs}}{9\,\text{mon}} \cdot \frac{1\,\text{mon}}{30\,\text{day}} \cdot \frac{1\,\text{day}}{24\,\text{hr}} \cdot \frac{16\,\text{oz}}{1\,\text{lb}}$$
$$\approx 0.173\,\frac{\text{oz}}{\text{hr}}$$

20.
$$\frac{133\,\text{lb}}{18\,\text{yr}} \cdot \frac{1\,\text{yr}}{365\,\text{day}} \cdot \frac{1\,\text{day}}{24\,\text{hr}} \cdot \frac{16\,\text{oz}}{1\,\text{lb}}$$
$$\approx 0.0135\,\frac{\text{oz}}{\text{hr}}$$

21.
$$d = \frac{45^2}{30 \cdot 0.6} = \frac{2025}{18} = 112.5\,\text{ft}$$

22.
$$208 = \frac{s^2}{30 \cdot 0.4} = \frac{s^2}{12}$$
$$s^2 = 12 \cdot 208 = 2496$$
$$s = \sqrt{2496} \approx 50\,\text{mph}$$

23. $V = \dfrac{4}{3}\pi r^3$, Volume varies directly with the cube of the radius.
$$k = \frac{4}{3}\pi$$

24. $\dfrac{4}{6} = \dfrac{EF}{15}$, $6EF = 60$, $EF = 10\,\text{units}$
$$AB = \sqrt{6^2 + 15^2} = \sqrt{261} = 3\sqrt{29}$$
$$\approx 16.16\,\text{units}$$
$$DE = \sqrt{4^2 + 10^2} = \sqrt{116} = 2\sqrt{29}$$
$$\approx 10.77\,\text{units}$$

25. a) Same distance, inverse proportion.

    b) $k = 60 \cdot 6 = 360\,\text{miles}$

    c) $60 \cdot 6 = t \cdot 55$, $t \approx 6.545\,\text{hrs}$

26.
$$\frac{x - 2}{10} = \frac{2}{x - 1},\ (x - 2)\,(x - 1) = 20$$
$$x^2 - 3x + 2 = 20,\quad x^2 - 3x - 18 = 0$$
$$(x - 6)\,(x + 3) = 0,\quad x - 6 = 0,\quad x = 6$$
$$x + 3 = 0,\quad x = -3$$
$$\{-3, 6\}\,;\ x \neq 1$$

27.
$$\frac{1}{3} - \frac{1}{x} = \frac{2}{3}x\,;\ x \neq 0$$
$$3x\left(\frac{1}{3} - \frac{1}{x}\right) = \frac{2}{3}x \cdot 3x$$
$$x - 3 = 2x^2$$
$$2x^2 - x + 3 = 0$$
$$x = \frac{-(-1) \pm \sqrt{(-1)^2 - 4(2)\,(3)}}{2(2)}$$
$$x = \frac{1 \pm \sqrt{-23}}{4} = \frac{1}{4} \pm i\,\frac{\sqrt{23}}{4}$$
No real solutions.
$$\left\{\frac{1}{4} \pm i\,\frac{\sqrt{23}}{4}\right\}$$

Chapter 4 Test

28. $\dfrac{1}{x-4} + \dfrac{1}{x} = \dfrac{10}{3x}$; $x \neq 0, 4$

$3x(x-4)\left(\dfrac{1}{x-4} + \dfrac{1}{x}\right) = \dfrac{10}{3x} \cdot 3x(x-4)$

$3x + 3x - 12 = 10x - 40$

$6x - 12 = 10x - 40$

$4x = 28$

$x = 7$

29. $\dfrac{x-1}{9} = \dfrac{x+3}{10}$

$10x - 10 = 9x + 27$

$x = 37$

30. $\dfrac{3}{x} + \dfrac{2}{x+1} = 4$; $x \neq -1, 0$

$x(x+1)\left(\dfrac{3}{x} + \dfrac{2}{x+1}\right) = 4 \cdot x \cdot (x+1)$

$3x + 3 + 2x = 4x^2 + 4x$

$5x + 3 = 4x^2 + 4x$

$4x^2 - x - 3 = 0$

$(4x + 3)(x - 1) = 0$

$4x + 3 = 0, \ x = -\dfrac{3}{4}$

$x - 1 = 0, \ x = 1$

$\left\{-\dfrac{3}{4}, \ 1\right\}$

31. $\dfrac{x^3 + 6x^2 + 12x + 8}{x + 2}$; defined for all $\mathbb{R}$ except $x = -2$

$$
\begin{array}{r}
x^2 + 4x + 4 \\
x + 2 \overline{\smash{\big)}\ x^3 + 6x^2 + 12x + 8} \\
\underline{-(x^3 + 2x^2)} \\
4x^2 + 12x \\
\underline{-(4x^2 + 8x)} \\
4x + 8 \\
\underline{-(4x + 8)} \\
0
\end{array}
$$

$x^2 + 4x + 4$

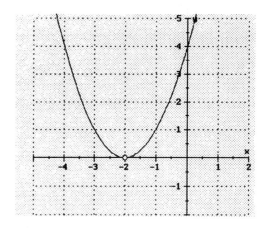

32. $\dfrac{x^3 + 6x^2 + 12x + 8}{x + 4}$; defined for all $\mathbb{R}$ except $x = -4$

$$
\begin{array}{r}
x^2 + 2x + 4 \\
x + 4 \overline{\smash{\big)}\ x^3 + 6x^2 + 12x + 8} \\
\underline{-(x^3 + 4x^2)} \\
2x^2 + 12x \\
\underline{-(2x^2 + 8x)} \\
4x + 8 \\
\underline{-(4x + 16)} \\
- 8
\end{array}
$$

$x^2 + 2x + 4 - \dfrac{8}{x + 4}$

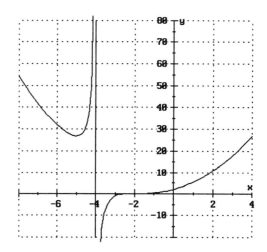

Section 5.0

1. a) $2x^{-1} = \dfrac{2}{x}$

    b) $\left(\dfrac{y}{x}\right)^{-1} = \dfrac{y^{-1}}{x^{-1}} = \dfrac{\frac{1}{y}}{\frac{1}{x}} = \dfrac{1}{y} \cdot \dfrac{x}{1} = \dfrac{x}{y}$

    c) $\left(\dfrac{S}{t}\right)^0 = 1$

    d) $0.25^{-1} = \left(\dfrac{25}{100}\right)^{-1} = \left(\dfrac{1}{4}\right)^{-1} = \dfrac{1^{-1}}{4^{-1}}$

    $= \dfrac{1}{\frac{1}{4}} = 1 \cdot \dfrac{4}{1} = 4$

    e) $\left(\dfrac{1}{2}\right)^0 = 1$

3. a) $\left(\dfrac{m}{n}\right)^{-1} = \dfrac{m^{-1}}{n^{-1}} = \dfrac{\frac{1}{m}}{\frac{1}{n}} = \dfrac{1}{m} \cdot \dfrac{n}{1} = \dfrac{n}{m}$

    b) $\left(\dfrac{b}{c}\right)^0 = 1$

    c) $\left(\dfrac{c}{ab}\right)^{-1} = \dfrac{c^{-1}}{(ab)^{-1}} = \dfrac{\frac{1}{c}}{\frac{1}{ab}}$

    $= \dfrac{1}{c} \cdot \dfrac{ab}{1} = \dfrac{ab}{c}$

    d) $\left(\dfrac{3}{4}\right)^0 = 1$

    e) $2.5^{-1} = \dfrac{1}{2.5} = \dfrac{10}{25} = \dfrac{2}{5} = 0.4$

5. a) $(2y)^{-2} = \dfrac{1}{(2y)^2} = \dfrac{1}{4y^2}$

5. b) $2x^{-4} = \dfrac{2}{x^4}$

    c) $\left(\dfrac{2x}{y}\right)^{-2} = \left(\dfrac{y}{2x}\right)^2 = \dfrac{y^2}{4x^2}$

    d) $\left(\dfrac{a}{2c}\right)^{-3} = \left(\dfrac{2c}{a}\right)^3 = \dfrac{8c^3}{a^3}$

    e) $\dfrac{2x^3 y^{-2}}{6x^{-1} y^2} = \dfrac{2x^3 x^1}{6y^2 y^2} = \dfrac{x^{3+1}}{3y^{2+2}} = \dfrac{x^4}{3y^4}$

    f) $b^1 \cdot b^n = b^{1+n} = b^{n+1}$

7. a) $\left(\dfrac{3a^2}{c}\right)^{-3} = \left(\dfrac{c}{3a^2}\right)^3 = \dfrac{c^3}{27a^6}$

    b) $\left(\dfrac{a}{c^2}\right)^{-2} = \left(\dfrac{c^2}{a}\right)^2 = \dfrac{c^4}{a^2}$

    c) $\dfrac{1}{c^{-3}} = c^3$

    d) $\dfrac{2}{a^{-2}} = 2a^2$

    e) $\dfrac{12a^3 b^{-2}}{4a^{-1} b^{-5}}$

    $= \dfrac{12}{4} \cdot a^3 \cdot a^1 \cdot b^5 \cdot b^{-2} = 3a^4 b^3$

    f) $3 \cdot 3^x = 3^{1+x} = 3^{x+1}$

9. The -2 exponent applies to **x** only.

11. a) $\sqrt{0.4} = \sqrt{0.04 \cdot 10}$

    $= 0.2\sqrt{10} \approx 0.632$

    b) $\sqrt{4} = \sqrt{2^2} = 2$

11. c) $\sqrt{40} = \sqrt{4 \cdot 10} = \sqrt{4} \cdot \sqrt{10}$
$= 2\sqrt{10} \approx 6.325$

d) $\sqrt{400} = \sqrt{4 \cdot 100} = \sqrt{4} \cdot \sqrt{100}$
$= 2 \cdot 10 = 20$

e) $\sqrt{4000} = \sqrt{4 \cdot 1000}$
$= \sqrt{4} \cdot \sqrt{100 \cdot 10} = 2 \cdot 10\sqrt{10}$
$= 20\sqrt{10} \approx 63.246$

13. a) $\sqrt{400,000} = \sqrt{4 \cdot 10^5}$
$= \sqrt{4 \cdot 10^4 \cdot 10} = 2 \cdot 10^2\sqrt{10}$
$= 200\sqrt{10} \approx 632.456$

b) $\sqrt{4,000,000} = \sqrt{4 \cdot 10^6}$
$= 2 \cdot 10^3 = 2000$

c) $\sqrt{0.04} = \sqrt{4 \cdot 10^{-2}}$
$= 2 \cdot 10^{-1} = 0.2$

d) $\sqrt{0.000004} = \sqrt{4 \cdot 10^{-6}}$
$= 2 \cdot 10^{-3} = 0.002$

15. Answers will vary-examples are:
$\sqrt{2,500}$ and $\sqrt{0.0025}$

17. Answers will vary-examples are:
$\sqrt{0.08}$ and $\sqrt{80,000}$

19. $\sqrt{100n} = \sqrt{100} \cdot \sqrt{n} = 10\sqrt{n}$

21. $\sqrt{0.01n} = \sqrt{0.01} \cdot \sqrt{n} = 0.1\sqrt{n}$

23. $\sqrt{\dfrac{n}{100}} = \dfrac{\sqrt{n}}{\sqrt{100}} = \dfrac{\sqrt{n}}{10}$

25. All those with the same digits have radicands containing 8 times an _even_ power of 10. 80 and 8 are 8 times an _odd_ power of 10.

27. $\Delta x = 6 - 2 = 4$
$\Delta y = -2 - 5 = -7$
Slope $= \dfrac{\Delta y}{\Delta x} = -\dfrac{7}{4}$
Distance $= \sqrt{\Delta x^2 + \Delta y^2}$
$= \sqrt{4^2 + (-7)^2}$
$= \sqrt{16 + 49} = \sqrt{65} \approx 8.062$

29. $\Delta x = 4 - (-1) = 5$
$\Delta y = -3 - 5 = -8$
Slope $= \dfrac{\Delta y}{\Delta x} = \dfrac{-8}{5}$
Distance $= \sqrt{\Delta x^2 + \Delta y^2}$
$= \sqrt{5^2 + (-8)^2}$
$= \sqrt{25 + 64} = \sqrt{89} = 9.434$

31. $\Delta x = c - a$
$\Delta y = d - b$
Slope $= \dfrac{\Delta y}{\Delta x} = \dfrac{d - b}{c - a}$
$= \dfrac{-1(b - d)}{-1(a - c)} = \dfrac{b - d}{a - c}$
Distance $= \sqrt{\Delta x^2 + \Delta y^2}$
$= \sqrt{(c - a)^2 + (d - b)^2}$

33. Set $A = \left\{ x^2, 3x^2, \dfrac{1}{2}x^3, x^4, x^{\frac{1}{3}}, \dfrac{1}{4}x^{\frac{1}{2}}, x^0, x^{-2}, x^{\pi} \right\}$

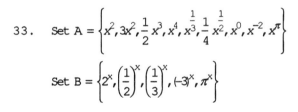

Set $B = \left\{ 2^x, \left(\dfrac{1}{2}\right)^x, \left(\dfrac{1}{3}\right)^x, (-3)^x, \pi^x \right\}$

37.

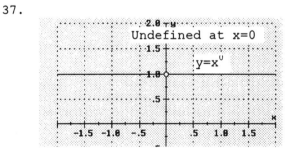

39. Answers will vary.
    Draw a right triangle with sides
    of 1 and 2, hypotenuse will be
    $\sqrt{5}$.

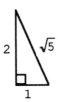

For other values some examples
are:

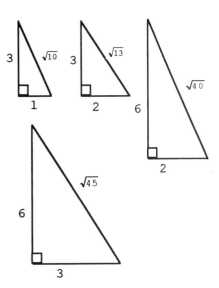

1. a) $\left(\dfrac{4}{25}\right)^{1/2} = \dfrac{4^{1/2}}{25^{1/2}} = \dfrac{2}{5}$

   b) $\left(\dfrac{1}{8}\right)^{1/3} = \dfrac{1}{8^{1/3}} = \dfrac{1}{2}$

   c) $\left(\dfrac{64}{27}\right)^{2/3} = \dfrac{64^{2/3}}{27^{2/3}}$

   $= \dfrac{\left(64^{1/3}\right)^2}{\left(27^{1/3}\right)^2} = \dfrac{4^2}{3^2} = \dfrac{16}{9}$

   d) $\left((64)^{1/2}\right)^{1/3} = (8)^{1/3} = 2$

3. a) $32^{2/5} = \left((32)^{1/5}\right)^2 = 2^2 = 4$

   b) $27^{2/3} = \left((27)^{1/3}\right)^2 = 3^2 = 9$

   c) $\left(\dfrac{1}{64}\right)^{-1/6} = (64)^{1/6} = 2$

   d) $4^{5/2} = \left(4^{1/2}\right)^5 = 2^5 = 32$

5. a) $8^{-1/3} = \left(8^{1/3}\right)^{-1} = 2^{-1} = \dfrac{1}{2}$

   b) $\left(\dfrac{3}{4}\right)^{-2} = \left(\dfrac{4}{3}\right)^2 = \dfrac{16}{9}$

   c) $\left(\dfrac{1}{3}\right)^{-2} = \left(\dfrac{3}{1}\right)^2 = 9$

   d) $32^{4/5} = \left(32^{1/5}\right)^4 = 2^4 = 16$

7. $1b = 5a,\ 1d = 3c$

   $b^{-n} = \left(\dfrac{1}{b}\right)^n$

9. a) $16^{1.5} = 16^{3/2} = \left(16^{1/2}\right)^3$

   $= 4^3 = 64$

9. b) $36^{1.5} = 36^{3/2} = \left(36^{1/2}\right)^3$

   $= 6^3 = 216$

   c) $9^{2.5} = 9^{2+0.5} = 9^2 \cdot 9^{0.5}$

   $= 81 \cdot 9^{1/2} = 81 \cdot 3 = 243$

   or $9^{2.5} = 9^{5/2} = \left(9^{1/2}\right)^5$

   $= 3^5 = 243$

   d) $0.01^{0.5} = 0.01^{1/2} = 0.1$

11. a) $\sqrt[3]{64} = 64^{1/3} = 4$

    b) $\sqrt[5]{32} = 32^{1/5} = 2$

    c) $\sqrt[3]{-125} = (-125)^{1/3} = -5$

    d) $\sqrt[7]{-128} = (-128)^{1/7} = -2$

13. When $x \geq 0$

15. a) $\sqrt{y} = |y|$

    b) $\sqrt{z^8} = z^4$

    c) $\sqrt{x^6} = |x^3|$

17. Answers will vary - examples are:
    $a = b = 2$ or $a = 2,\ b = 4$

19. True for any $b > 0$, except for $b = 1$.

21. a) $4^x = 256,\ x = 4$
    b) $4^x = 500,\ x \approx 4.5$ Estimated from graph
    c) $4^x = 1000,\ x \approx 5$

23. a) Annual
    b) Semiannual
    c) Quarterly
    d) Monthly
    e) Weekly
    f) Daily

25. Base $= 1 + \dfrac{r}{n}$

27. a) $10000\left(1 + \dfrac{0.04}{12}\right)^{(12 \cdot 3)} = 11{,}272.72$

    b) $10000\left(1 + \dfrac{0.04}{4}\right)^{(4 \cdot 3)} = 11{,}268.25$

29. a) $10000\left(1 + \dfrac{0.08}{12}\right)^{(12 \cdot 3)} = 12{,}702.37$

    b) $10000\left(1 + \dfrac{0.08}{4}\right)^{(4 \cdot 3)} = 12{,}682.42$

31. Answers will vary, example:
    $10,000 earning 4% compounded
    monthly (quarterly) for 3 years.

33. $6.00(1 + 0.05)^{10} = \$9.77$

35. $1000\left(1 + \dfrac{0.07}{4}\right)^{(4 \cdot 5)} = \$1{,}414.78$

37. $x^3 = 10,\ x = \sqrt[3]{10} \approx 2.154$

    $x^4 = 10,\ x = \sqrt[4]{10} \approx 1.778$

    $x^5 = 10,\ x = \sqrt[5]{10} \approx 1.585$

    $x^6 = 10,\ x = \sqrt[6]{10} \approx 1.468$

39. Base x is approaching 1.

1. a) $16^{3/4} = \left(16^{1/4}\right)^3 = 2^3 = 8$

   b) $64^{3/2} = \left(64^{1/2}\right)^3 = 8^3 = 512$

   c) $0.09^{3/2} = \left(0.09^{1/2}\right)^3 = (0.3)^3 = 0.027$

3. a) $81^{3/4} = \left(81^{1/4}\right)^3 = 3^3 = 27$

   b) $27^{4/3} = \left(27^{1/3}\right)^4 = 3^4 = 81$

   c) $0.008^{2/3} = \left(0.008^{1/3}\right)^2 = (0.2)^2 = 0.04$

5. a) $\sqrt{2}\sqrt{18} = \sqrt{2 \cdot 18} = \sqrt{36} = 6$

   b) $-\sqrt{3}\sqrt{27} = -\sqrt{3 \cdot 27} = -\sqrt{81} = -9$

   c) $\sqrt{2}\sqrt{8} = \sqrt{2 \cdot 8} = \sqrt{16} = 4$

   d) $\sqrt[2]{\sqrt[3]{64}} = \sqrt[2]{4} = 2$

7. The even root of a negative number is not a real number.

9. "The real number"

11. For all real numbers.

13. $A = \dfrac{\pi r^2 \theta}{360}$, $A = \dfrac{\pi(5\ \text{in})^2(44)}{360}$

    $A \approx 9.599\ \text{in}^2$

15. $22\ \text{in}^2 = \dfrac{\pi r^2 (20)}{360}$, $r^2 = \dfrac{360 \cdot 22\ \text{in}^2}{\pi \cdot 20}$

    $r = \sqrt{\left(\dfrac{18 \cdot 22\ \text{in}^2}{\pi}\right)} \approx 11.227\ \text{in.}$

17. $v = \dfrac{4}{3}\pi r^3$, $r^3 = \dfrac{3v}{4\pi}$

    $r = \sqrt[3]{\dfrac{3v}{4\pi}}$

19. $v = \dfrac{\pi p r^4}{8Ln}$, $r^4 = \dfrac{8Lnv}{\pi p}$

    $r = \sqrt[4]{\dfrac{8Lnv}{\pi p}}$

21. a) $\sqrt{a}\sqrt{a^3} = \sqrt{a^4} = a^2;\ a \geq 0$

    b) $\sqrt[3]{b}\sqrt[3]{b^2} = \sqrt[3]{b^3} = b$

23. a) $x^{3/4}x^{1/3} = x^{3/4+1/3} = x^{9/12+4/12}$

       $= x^{13/12}$

    b) $x^{3/4}x^{2/3} = x^{3/4+2/3} = x^{9/12+8/12}$

       $= x^{17/12}$

    c) $x^{-1/3} = \dfrac{1}{x^{1/3}}$

    d) $\sqrt[3]{x^2}\sqrt[4]{x} = \left(x^2\right)^{1/3}(x)^{1/4} = x^{2/3} \cdot x^{1/4}$

       $= x^{2/3+1/4} = x^{8/12+3/12} = x^{11/12}$

25. a) $\sqrt[3]{x}\sqrt[2]{x^3} = (x)^{1/3}\left(x^3\right)^{1/2} = x^{1/3} \cdot x^{3/2}$

       $= x^{1/3+3/2} = x^{2/6+9/6} = x^{11/6}$

    b) $\dfrac{x^{3/4}}{x^{1/3}} = (x)^{3/4} \cdot (x)^{-1/3} = x^{3/4-1/3}$

       $= x^{9/12-4/12} = x^{5/12}$

    c) $\left(x^{3/4}\right)^{2/3} = (x)^{3/4 \cdot 2/3} = x^{6/12}$

       $= x^{1/2}$

27. a) $\sqrt[2]{\sqrt[3]{x^2}} = \left(\left(x^2\right)^{1/3}\right)^{1/2} = \left(x^{2/3}\right)^{1/2}$

       $= x^{2/3 \cdot 1/2} = x^{1/3}$

    b) $\left(\dfrac{16x^4}{y^8}\right)^{3/4} = \dfrac{16^{3/4}x^{4 \cdot 3/4}}{y^{8 \cdot 3/4}} = \dfrac{\left(16^{1/4}\right)^3 x^3}{y^6}$

       $= \dfrac{2^3 x^3}{y^6} = \dfrac{8x^3}{y^6}$

Section 5.2

27. c) $\dfrac{x^{-1/2}}{x^{3/2}} = x^{-1/2} \cdot x^{-3/2} = x^{-1/2-3/2}$

$\qquad = x^{-4/2} = x^{-2} = \dfrac{1}{x^2}$

29. Yes; $-x$ is > 0 if $x < 0$

31. 3 is not a factor of $\sqrt{21}$ and the numerator must be factored before any canceling with the denominator.

33. Window may vary
$xmin = 9.8866$, $ymin = 1989.9844$
$xmax = 9.9546$, $ymax = 2006.3736$
$t \approx 9.9088$

35. a)

Input	Output
1970	14,000
1980	28,000
1990	56,000
2000	112,000
2010	224,000
2020	448,000
2030	896,000

b) 75,000 is between 56,000 (1990) and 112,000 (2000).

c) $14,000(1+0.07)^{40} \approx \$209{,}642$, rounded to $\approx \$210{,}000$.

37. $5 = 96(1 + r)^{21}$

$\dfrac{5}{96} = (1 + r)^{21}$

$\left(\dfrac{5}{96}\right)^{1/21} = 1 + r$

$r = \left(\dfrac{5}{96}\right)^{1/21} - 1 \approx -0.13$

Inflation rate $\approx -13\%$

39. $100{,}000(1 + 0.05)^{30} \approx \$432{,}000$

41. $3000 = 40{,}000(1 + r)^{10}$

$\dfrac{3000}{40{,}000} = (1 + r)^{10} \quad \dfrac{3}{40} = (1 + r)^{10}$

$\left(\dfrac{3}{40}\right)^{1/10} = 1 + r \quad \left(\dfrac{3}{10}\right)^{1/10} - 1 = r$

$r \approx -0.228$

Depreciation rate $= 22.8\%$

43. From $\left(1 + \dfrac{r}{4}\right)$ to $\left(1 + \dfrac{r}{12}\right)$

45. a) $\dfrac{100\,\text{lbs}}{2.205\,\dfrac{\text{lbs}}{\text{kg}}} \approx 45.351\,\text{kg}$

$\dfrac{130\,\text{lbs}}{2.205\,\dfrac{\text{lbs}}{\text{kg}}} \approx 58.957\,\text{kg}$

$5'1'' = 61''; \quad 61'' \cdot 2.54\,\dfrac{\text{cm}}{\text{in}} \approx 154.94\,\text{cm}$

$\dfrac{(45.351)^{1.2}}{(154.94)^{3.3}} \cdot 4{,}000{,}000 \approx 23.04$

$\dfrac{(58.957)^{1.2}}{(154.94)^{3.3}} \cdot 4{,}000{,}000 \approx 31.56$

Range 23.04 to 31.56

b)

$\dfrac{\left(\dfrac{132}{2.205}\right)^{1.2}}{(70 \cdot 2.54)^{3.3}} \cdot 4{,}000{,}000 \approx 20.41$

$\dfrac{\left(\dfrac{165}{2.205}\right)^{1.2}}{(70 \cdot 2.54)^{3.3}} \cdot 4{,}000{,}000 \approx 26.68$

Range 20.41 to 26.68

c) $\dfrac{\left(\dfrac{153}{2.205}\right)^{1.2}}{(74 \cdot 2.54)^{3.3}} \cdot 3{,}000{,}000 \approx 15.21$

$\dfrac{\left(\dfrac{193}{2.205}\right)^{1.2}}{(74 \cdot 2.54)^{3.3}} \cdot 3{,}000{,}000 \approx 20.10$

Range 15.21 to 20.10

47. $(2i)(2i) = 2 \cdot 2 \cdot i \cdot i = 4i^2 = 4 \cdot (-1) = -4$

49.

	$\sqrt{2}$	$-i\sqrt{2}$
$\sqrt{2}$	2	$-2i$
$-i\sqrt{2}$	$-2i$	$+i^2 \cdot 2 = -2$

$$\left(\sqrt{2} - i\sqrt{2}\right)\left(\sqrt{2} - i\sqrt{2}\right) = -4i$$

$$(-4i)(-4i) = 16i^2 = -16$$

51. $5 \cdot 5^n = 5^1 \cdot 5^n = 5^{1+n}$

We add the exponents when the bases are the same; we do **not** multiply the bases.

Mid-Chapter 5 Test

1. $\sqrt{x}$ Other answers are possible.

2. x^{-1} Other answers are possible.

3. $\dfrac{a^{-1}b^2}{c^0} = \dfrac{b^2}{a}$

4. $\dfrac{a^2 b^{-1}}{c^{-2}} = \dfrac{a^2 c^2}{b}$

5. $a \cdot a^x = a^1 \cdot a^x = a^{x+1}$

6.

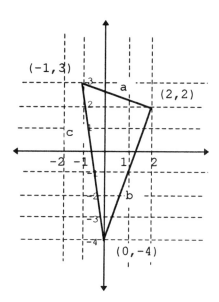

(−1, 3)   (2, 2)   (0, −4)

Length of sides.

$a = \sqrt{(3-2)^2 + (-1-2)^2}$
$\quad = \sqrt{1+9} = \sqrt{10}$

$b = \sqrt{(2-0)^2 + (2-(-4))^2}$
$\quad = \sqrt{4+36} = 2\sqrt{10}$

$c = \sqrt{(-1-0)^2 + (3-(-4))^2}$
$\quad = \sqrt{1+49} = 5\sqrt{2}$

$a^2 + b^2 = \left(\sqrt{10}\right)^2 + \left(2\sqrt{10}\right)^2$
$\quad = 10 + 40 = 50$

$c^2 = \left(\sqrt{50}\right)^2 = 50$

$a^2 + b^2 = c^2,$

This is a right triangle.

7. a) $\sqrt{4n} = 2\sqrt{n}, \ n \geq 0$

   b) $\sqrt{40n^2} = 2|n|\sqrt{10}$

   c) $\sqrt{400n} = 20\sqrt{n}, \ n \geq 0$

   d) $\sqrt{4000n} = 20\sqrt{10n}, \ n \geq 0$

8. $\$1000\left(1 + \dfrac{0.07}{52}\right)^{(52 \cdot 5)} \approx \$1418.73$

9. $\$2500 = \$1500(1 + r)^6$

   $\dfrac{2500}{1500} = (1+r)^6 \qquad \left(\dfrac{5}{3}\right)^{1/6} - 1 = r$

   $r \approx 0.0889$

   Annual interest rate $\approx 8.89\%$

10. $T = 2\pi\sqrt{\dfrac{L}{g}}$

    $\dfrac{T}{2\pi} = \sqrt{\dfrac{L}{g}} \qquad \left(\dfrac{T}{2\pi}\right)^2 = \dfrac{L}{g}$

    $g = \dfrac{L}{\dfrac{T^2}{4\pi^2}} = \dfrac{4\pi^2 L}{T^2}$

11. a) $x^3 + y^3 = a^3 \quad y^3 = a^3 - x^3$

    $y = \sqrt[3]{a^3 - x^3}$

    b) $x^2 = \dfrac{8}{27k}(y-k)^3$

    $(y-k)^3 = \dfrac{27kx^2}{8}$

    $y - k = \left(\dfrac{27kx^2}{8}\right)^{1/3} = \dfrac{3k^{1/3}x^{2/3}}{2}$

    $y = \dfrac{3k^{1/3}x^{2/3}}{2} + k = \dfrac{3}{2}\sqrt[3]{kx^2} + k$

12. a) $8^{2/3} = \left(\sqrt[3]{8}\right)^2 = 2^2 = 4$

    b) $125^{1/3} = \sqrt[3]{125} = 5$

13. a) $16^{3/4} = \left(\sqrt[4]{16}\right)^3 = 2^3 = 8$

    b) $\left(a^3\right)^{2/3} = \left(\sqrt[3]{a^3}\right)^2 = a^2$

14. a) $\sqrt[3]{-64} = -4$

    b) $\sqrt[4]{16y^4} = 2|y|$

15. a) $4^x = 64 = 4^3$
       $x = 3$

    b) $3^x = 27 = 3^3$
       $x = 3$

    c) $\dfrac{1}{2^x} = \dfrac{1}{8} = \dfrac{1}{2^3}$
       $x = 3$

    d) $\pi^x = 1 = \pi^0$
       $x = 0$

1. a) $\left(2 - \sqrt{3}\right)\left(2 + \sqrt{3}\right) = 2^2 - \left(\sqrt{3}\right)^2$

    $= 4 - 3 = 1$

    b) $\left(5 - \sqrt{2}\right)\left(5 - \sqrt{2}\right)$

    $= 5^2 - 10\sqrt{2} + \left(\sqrt{2}\right)^2$

    $= 25 - 10\sqrt{2} + 2 = 27 - 10\sqrt{2}$

3. a) $\left(x - \sqrt{5}\right)\left(x + \sqrt{5}\right)$

    $= x^2 - \left(\sqrt{5}\right)^2$

    $= x^2 - 5$

    b) $\left(x - \sqrt{7}\right)\left(x - \sqrt{7}\right)$

    $= x^2 - 2\sqrt{7}x + \left(-\sqrt{7}\right)^2$

    $= x^2 - 2\sqrt{7} + 7$

5. a) $\sqrt{3}\sqrt{12} = \sqrt{3 \cdot 12} = \sqrt{36} = 6$

    b) $\sqrt{-3}\sqrt{12} = \sqrt{-3 \cdot 12} = \sqrt{-36} = 6i$

7. a) $\sqrt{-125}\sqrt{-5} = \sqrt{-125 \cdot (-5)} = \sqrt{625} = 25$

    b) $\sqrt{5}\sqrt{-125} = \sqrt{5 \cdot (-125)} = \sqrt{-625} = 25i$

9. $3\sqrt{2} - 4\sqrt{2} + 7\sqrt{2} = (3 - 4 + 7)\sqrt{2} = 6\sqrt{2}$

11. $\sqrt{20} + \sqrt{45} = \sqrt{4 \cdot 5} + \sqrt{9 \cdot 5}$

    $= 2\sqrt{5} + 3\sqrt{5} = 5\sqrt{5}$

13. $\sqrt{9x} + \sqrt{16x} - \sqrt{x} = 3\sqrt{x} + 4\sqrt{x} - \sqrt{x}$

    $= (3 + 4 - 1)\sqrt{x} = 6\sqrt{x}$

15. $\sqrt{0.01x} + \sqrt{49x} - \sqrt{16x} = 0.1\sqrt{x} + 7\sqrt{x} - 4\sqrt{x}$

    $= (0.1 + 7 - 4)\sqrt{x} = 3.1\sqrt{x}$

17. a) $\dfrac{\sqrt[4]{243}}{\sqrt[4]{3}} = \sqrt[4]{\dfrac{243}{3}} = \sqrt[4]{81} = 3$

    b) $\sqrt[3]{16} \cdot \sqrt[3]{4} = \sqrt[3]{16 \cdot 4} = \sqrt[3]{64} = 4$

19. a) $\left(\sqrt{x} + 3\right)^2 = \left(\sqrt{x}\right)^2 + 6\sqrt{x} + 3^2$

    $= x + 6\sqrt{x} + 9$

    b) $\left(\sqrt{a} - 3\right)^2 = \left(\sqrt{a}\right)^2 - 6\sqrt{a} + (-3)^2$

    $= a - 6\sqrt{a} + 9$

    c) $\left(2 - \sqrt{a}\right)^2 = 2^2 - 4\sqrt{a} + (-\sqrt{a})^2$

    $= 4 - 4\sqrt{a} + a$

    d) $(\sqrt{a} - \sqrt{b})(\sqrt{a} + \sqrt{b}) = \left(\sqrt{a}\right)^2 - \left(\sqrt{b}\right)^2$

    $= a - b$

21. a. $3 - \sqrt{2}$ or $-3 + \sqrt{2}$

    b) $\sqrt{a} + 3$ or $-\sqrt{a} - 3$

    c) $a + \sqrt{b}$

    d) $3 + 2i$

    e) $2 - 7i$

23. a) $\dfrac{4}{\sqrt{5}} \cdot \dfrac{\sqrt{5}}{\sqrt{5}} = \dfrac{4\sqrt{5}}{5}$

    b) $\dfrac{8}{\sqrt{6}} \cdot \dfrac{\sqrt{6}}{\sqrt{6}} = \dfrac{8\sqrt{6}}{6} = \dfrac{4\sqrt{6}}{3}$

25. a) $\dfrac{a}{\sqrt{c}} \cdot \dfrac{\sqrt{c}}{\sqrt{c}} = \dfrac{a\sqrt{c}}{c}$

    b) $\dfrac{a}{\sqrt{a}} \cdot \dfrac{\sqrt{a}}{\sqrt{a}} = \dfrac{a\sqrt{a}}{a} = \sqrt{a}$

27. $\dfrac{4}{7 - \sqrt{5}} \cdot \dfrac{7 + \sqrt{5}}{7 + \sqrt{5}} = \dfrac{4\left(7 + \sqrt{5}\right)}{49 - 5}$

    $= \dfrac{4\left(7 + \sqrt{5}\right)}{44} = \dfrac{7 + \sqrt{5}}{11}$

29. $\dfrac{x}{x - \sqrt{y}} \cdot \dfrac{x + \sqrt{y}}{x + \sqrt{y}} = \dfrac{x\left(x + \sqrt{y}\right)}{x^2 - y}$

$= \dfrac{x^2 - x\sqrt{y}}{x^2 - y}$

31. a) $\dfrac{1}{3 + i} \cdot \dfrac{3 - i}{3 - i} = \dfrac{3 - i}{9 - (-1)} = \dfrac{3 - i}{10}$

b) $\dfrac{i}{3 - i} \cdot \dfrac{3 + i}{3 + i} = \dfrac{i(3 + i)}{9 - (-1)}$

$= \dfrac{3i + (-1)}{10} = \dfrac{-1 + 3i}{10}$

c) $\dfrac{2 - i}{1 + 2i} \cdot \dfrac{1 - 2i}{1 - 2i} = \dfrac{(2 - i)(1 - 2i)}{1 - (-4)}$

$= \dfrac{2 - i - 4i + 2i^2}{5} = \dfrac{2 - 5i - 2}{5}$

$= \dfrac{-5i}{5} = -i$

33. $\left(-2 - \sqrt{2}\right)^2 + 4\left(-2 - \sqrt{2}\right) + 2 \overset{?}{=} 0$

$4 + 4\sqrt{2} + 2 - 8 - 4\sqrt{2} + 2 \overset{?}{=} 0$

$4 + 2 - 8 + 2 + 4\sqrt{2} - 4\sqrt{2} \overset{?}{=} 0$

$0 = 0$

35. $\left(-1 + i\sqrt{3}\right)^2 + 2\left(-1 + i\sqrt{3}\right) + 4 \overset{?}{=} 0$

$1 - 2i\sqrt{3} + 3i^2 - 2 + 2i\sqrt{3} + 4 \overset{?}{=} 0$

$1 - 2i\sqrt{3} - 3 - 2 + 2i\sqrt{3} + 4 \overset{?}{=} 0$

$1 - 3 - 2 + 4 - 2i\sqrt{3} + 2i\sqrt{3} \overset{?}{=} 0$

$0 = 0$

37. a) $\left(\dfrac{1 + \sqrt{5}}{2}\right)\left(\dfrac{1 - \sqrt{5}}{2}\right) = \dfrac{1 - 5}{4} = \dfrac{-4}{4} = -1$

b) $\dfrac{1 + \sqrt{5}}{2} \approx 1.6180,\quad \dfrac{1 - \sqrt{5}}{2} \approx -0.6180$

c) Reciprocal of $\dfrac{1 + \sqrt{5}}{2}$ is $\dfrac{2}{1 + \sqrt{5}}$,

reciprocal of $\dfrac{1 - \sqrt{5}}{2} = \dfrac{2}{1 - \sqrt{5}}$

$\dfrac{2}{1 + \sqrt{5}} = -\dfrac{1 - \sqrt{5}}{2}$

& $\dfrac{2}{1 - \sqrt{5}} = -\dfrac{1 + \sqrt{5}}{2}$

39. $t = \dfrac{-b \pm \sqrt{b^2 - 4ac}}{2a} = \dfrac{1 \pm \sqrt{1 + 80}}{4}$

$a = 2,\ b = -1,\ c = -10$

$2t^2 - t - 10 = 0$

$\dfrac{1 + \sqrt{81}}{4} = \dfrac{1 + 9}{4} = \dfrac{10}{4} = \dfrac{5}{2}$

$\dfrac{1 - \sqrt{81}}{4} = \dfrac{1 - 9}{4} = \dfrac{-8}{4} = -2$

$t = \dfrac{5}{2}$ & $t = -2$

Two x intercepts

41. $x = \dfrac{10 \pm \sqrt{100 - 104}}{4}$

$a = 2,\ b = -10,\ c = 13$

$2x^2 - 10x + 13 = 0$

$\dfrac{10 + \sqrt{-4}}{4} = \dfrac{10 + 2i}{4} = \dfrac{5 + i}{2}$

$\dfrac{10 - \sqrt{-4}}{4} = \dfrac{10 - 2i}{4} = \dfrac{5 - i}{2}$

$x = \left\{\dfrac{5 - i}{2},\ \dfrac{5 + i}{2}\right\}$

No x - intercepts

43. $x = \dfrac{-2 \pm \sqrt{4 - 40}}{4}$

$a = 2,\ b = 2,\ c = 5$

$2x^2 + 2x + 5 = 0$

$\dfrac{-2 + \sqrt{-36}}{4} = \dfrac{-2 + 6i}{4} = \dfrac{-1 + 3i}{2}$

$\dfrac{-2 - \sqrt{-36}}{4} = \dfrac{-2 - 6i}{4} = \dfrac{-1 - 3i}{2}$

$x = \left\{ \dfrac{-1 - 3i}{2},\ \dfrac{-1 + 3i}{2} \right\}$

No x - intercepts

45. a) $\dfrac{1}{\sqrt[3]{2}} \cdot \dfrac{\sqrt[3]{2^2}}{\sqrt[3]{2^2}} = \dfrac{\sqrt[3]{2^2}}{\sqrt[3]{2^3}} = \dfrac{\sqrt[3]{4}}{2}$

b) $\dfrac{2}{\sqrt[3]{4}} = \dfrac{2}{\sqrt[3]{2^2}}$

$\dfrac{2}{\sqrt[3]{2^2}} \cdot \dfrac{\sqrt[3]{2}}{\sqrt[3]{2}} = \dfrac{2\sqrt[3]{2}}{\sqrt[3]{2^3}} = \dfrac{2\sqrt[3]{2}}{2} = \sqrt[3]{2}$

c) $\dfrac{3}{\sqrt[3]{3}} \cdot \dfrac{\sqrt[3]{3^2}}{\sqrt[3]{3^2}} = \dfrac{3\sqrt[3]{3^2}}{\sqrt[3]{3^3}}$

$= \dfrac{3\sqrt[3]{9}}{3} = \sqrt[3]{9}$

47. Slope BP $= \dfrac{y}{x + r}$

Substitute: $y = \sqrt{r^2 - x^2}$

Rationalize numerator:

$= \dfrac{\sqrt{r^2 - x^2}}{x + r} \cdot \dfrac{\sqrt{r^2 - x^2}}{\sqrt{r^2 - x^2}}$

Simplify: $= \dfrac{r^2 - x^2}{(x + r)\left(\sqrt{r^2 - x^2}\right)}$

Factor numerator:

$= \dfrac{(r + x)(r - x)}{(x + r)\left(\sqrt{r^2 - x^2}\right)}$

Simplify: $= \dfrac{r - x}{\sqrt{r^2 - x^2}}$

Simplify: $= \dfrac{(-1)(x - r)}{\sqrt{r^2 - x^2}}$

Substitute: $y = \sqrt{r^2 - x^2}$

$= \dfrac{(-1)(x - r)}{y} = (-1)\left(\dfrac{x - r}{y}\right)$

Reciprocal of AP is $\dfrac{x - r}{y}$;

this is the negative reciprocal of AP.

Section 5.4

1. a) $2^x = 256 = 2^8$

   $x = 8$

   b) $16^x = 256 = 2^8$

   $\left(2^4\right)^x = 2^8$

   $4x = 8$

   $x = 2$

3. a) $4^n = 8$

   $\left(2^2\right)^n = 2^3$

   $2n = 3$

   $n = \dfrac{3}{2}$

   b) $8^n = 16$

   $\left(2^3\right)^n = 2^4$

   $3n = 4$

   $n = \dfrac{4}{3}$

5. a) $16^n = 4$

   $\left(4^2\right)^n = 4^1$

   $2n = 1$

   $n = \dfrac{1}{2}$

   b) $100^x = 1000$

   $10^{2x} = 10^3$

   $2x = 3$

   $x = \dfrac{3}{2}$

7. a) $64^x = \dfrac{1}{8}$

   $8^{2x} = 8^{-1}$

   $2x = -1$

   $x = -\dfrac{1}{2}$

b) $125^n = \dfrac{1}{5}$

   $5^{3n} = 5^{-1}$

   $3n = -1$

   $n = -\dfrac{1}{3}$

9. a) $8^n = \dfrac{1}{2}$

   $2^{3n} = 2^{-1}$

   $3n = -1$

   $n = -\dfrac{1}{3}$

   b) $81^x = \dfrac{1}{3}$

   $3^{4x} = 3^{-1}$

   $4x = -1$

   $x = -\dfrac{1}{4}$

11. a) $\dfrac{1}{2^{x+5}} = 32$

   $2^{-(x+5)} = 2^5$

   $-x - 5 = 5$

   $-x = 10$

   $x = -10$

   b) $\left(\dfrac{1}{8}\right)^{x+3} = 8$

   $\left(8^{-1}\right)^{x+3} = 8$

   $-(x + 3) = 1$

   $x + 3 = -1$

   $x = -4$

13. a) $2^{x-1} = 2^{2x+5}$

   $x - 1 = 2x + 5$

   $x = -6$

   b) $3^{2x-1} = 3^{x+3}$

   $2x - 1 = x + 3$

   $x = 4$

15. a) $2^{x+5} = 4^{x-1}$

$2^{x+5} = (2^2)^{x-1}$

$x + 5 = 2(x - 1) = 2x - 2$

$x = 7$

b) $5^{x+1} = 125^{x-3}$

$5^{x+1} = (5^3)^{x-3}$

$x + 1 = 3(x - 3) = 3x - 9$

$2x = 10$

$x = 5$

17. Change to like base.

$p = 4^{6t} = \left(2^2\right)^{6t} = 2^{12t}$

$p = 8^{4t} = \left(2^3\right)^{4t} = 2^{12t}$

Results are the same.

19. From graph, x = 14.

21. From graph, no solution.

23. $\sqrt{2 - x} = 2 \quad x \geq 2$

$(2 - x) = 4$

$-x = 2$

$x = -2$

25. $\sqrt{(x - 5)} = 2, \quad x \geq 5$

$x - 5 = 4$

$x = 9$

27. $\sqrt{(x - 5)} - 6 = -2, \quad x \geq 5$

$\sqrt{(x - 5)} = 4$

$x - 5 = 16$

$x = 21$

29. $\sqrt{(2x + 14)} = 0, \quad x \geq -7$

$2x + 14 = 0$

$2x = -14$

$x = -7$

31. $\sqrt{(2x + 14)} = 6, \quad x \geq -7$

$2x + 14 = 36$

$2x = 22$

$x = 11$

33. $\sqrt{(3x - 5)} = 2, \quad x \geq \dfrac{5}{3}$

$3x - 5 = 4$

$3x = 9$

$x = 3$

35. $\sqrt{(3x - 5)} = -1, \quad x \geq \dfrac{5}{3}$

Principal square root is always positive, no solution.

37. $\sqrt{(5x - 1)} = 3, \quad x \geq \dfrac{1}{5}$

$5x - 1 = 9$

$5x = 10$

$x = 2$

39. $\sqrt{(3x - 5)} = \sqrt{2x} + 1, \quad x \geq \dfrac{5}{3}$

$3x - 5 = \left(\sqrt{2x} + 1\right)^2 = 2x + 2\sqrt{2x} + 1$

$x - 6 = 2\sqrt{2x}$

$(x - 6)^2 = 4 \cdot 2x = 8x$

$x^2 - 12x + 36 = 8x$

$x^2 - 20x + 36 = 0$

$(x - 18)(x - 2) = 0$

$x = 18 \text{ or } x = 2$

$\sqrt{3 \cdot 2 - 5} \overset{?}{=} \sqrt{2 \cdot 2} + 1$

$1 \overset{?}{=} 2 + 1 \quad \underline{\text{no.}}$

$\sqrt{3 \cdot 18 - 5} \overset{?}{=} \sqrt{2 \cdot 18} + 1$

$\sqrt{49} \overset{?}{=} \sqrt{36} + 1$

$7 \overset{?}{=} 6 + 1 \quad \underline{\text{yes}}$

$x = 18$

41. $\sqrt{(6x - 8)} = \sqrt{3x} + 2, \quad x \geq \dfrac{4}{3}$

$(6x - 8) = \left(\sqrt{3x} + 2\right)^2 = 3x + 4\sqrt{3x} + 4$

$3x - 12 = 4\sqrt{3x}$

$(3x - 12)^2 = 16 \cdot 3x = 48x$

$9x^2 - 72x + 144 = 48x$

$9x^2 - 120x + 144 = 0$

$(9x - 12)(x - 12) = 0$

$9x - 12 = 0 \quad or \quad x - 12 = 0$

$x = \dfrac{4}{3} \quad or \quad x = 12$

$\sqrt{6 \cdot \dfrac{4}{3} - 8} \overset{?}{=} \sqrt{3 \cdot \dfrac{4}{3}} + 2$

$0 \overset{?}{=} \sqrt{4} + 2$

$0 \overset{?}{=} 2 + 2 \quad \underline{no.}$

$\sqrt{6 \cdot 12 - 8} \overset{?}{=} \sqrt{3 \cdot 12} + 2$

$\sqrt{64} \overset{?}{=} \sqrt{36} + 2$

$8 \overset{?}{=} 6 + 2 \quad \underline{yes}$

$x = 12$

43. $\sqrt{(x + 1)} = \sqrt{(2x)} - 1, \quad x \geq 0$

$x + 1 = \left(\sqrt{(2x)} - 1\right)^2 = 2x - 2\sqrt{2x} + 1$

$x = 2\sqrt{2x}$

$x^2 = 4 \cdot 2x = 8x$

$x^2 - 8x = 0$

$x(x - 8) = 0$

$x = 0 \quad or \quad x = 8$

$\sqrt{0 + 1} \overset{?}{=} \sqrt{2 \cdot 0} - 1$

$1 \overset{?}{=} -1 \quad \underline{no.}$

$\sqrt{8 + 1} \overset{?}{=} \sqrt{2 \cdot 8} - 1$

$\sqrt{9} \overset{?}{=} \sqrt{16} - 1$

$3 \overset{?}{=} 4 - 1 \quad \underline{yes}$

$x = 8$

45. It is difficult to write 10 as a power of 2.

47. The principal square root is always positive.

49. In $y = \sqrt{x} - 7$ the $\sqrt{x}$ is evaluated then 7 is subracted from the result. In $y = \sqrt{x - 7}$ the 7 is subtracted from x then the square root is found.

51. $r = \sqrt{24L}$

$r^2 = 24L$

$L = \dfrac{r^2}{24}$

53. $r_{dry} = \sqrt{24 \cdot 50} \approx 34.64 \, mph$

$r_{wet} = \sqrt{12 \cdot 50} \approx 24.49 \, mph$

$\dfrac{r_{wet}}{r_{dry}} = \dfrac{\sqrt{12L}}{\sqrt{24L}} = \sqrt{\dfrac{12L}{24L}} = \sqrt{\dfrac{1}{2}}$

$= \dfrac{1}{\sqrt{2}} = \dfrac{\sqrt{2}}{2}$

55. $h = r - \sqrt{r^2 - \left(\dfrac{w}{2}\right)^2} \; ; \; r = 18, \; w = 33$

$h = 18 - \sqrt{18^2 - \left(\dfrac{33}{2}\right)^2}$

$h = 18 - \sqrt{18^2 - \dfrac{33^2}{4}} = 18 - \sqrt{51.75}$

$h \approx 10.806 \, ft$

57. $h = r - \sqrt{r^2 - \left(\dfrac{w}{2}\right)^2}$

$\sqrt{r^2 - \left(\dfrac{w}{2}\right)^2} = r - h$

$r^2 - \left(\dfrac{w}{2}\right)^2 = (r - h)^2 = r^2 - 2rh + h^2$

$r^2 - r^2 + 2rh - h^2 = \dfrac{w^2}{4}$

$4\left(2rh - h^2\right) = w^2$

$w = \sqrt{4(2rh - h^2)}$

$w = 2\sqrt{(2rh - h^2)}$

Section 5.5

1. a) $2^n = 0.1$

$2^{-3} = \frac{1}{8} = 0.125$

$2^{-4} = \frac{1}{16} = 0.0625$

$2^{-3.5} \approx 0.0884$

$2^{-3.3} \approx 0.1015$

$2^{-3.32} \approx 0.1001$

$n \approx -3.32$

b) $n \approx -2.32$

c) $n \approx -1.32$

d) $n \approx -0.32$

3. Starting with $2^{-3.32} = 0.1$

$0.2 = 2 \cdot 0.1 = 2^1 \cdot 2^{-3.32}$

$= 2^{1-3.32} = 2^{-2.32}$

0.4 is 4 times 0.1 or 2^2

so that m is changing but n

stays the same.

$0.4 = 2^2 \cdot 2^{-3.32} = 2^{2-3.32} = 2^{-1.32}$

$0.8 = 2^3 \cdot 2^{-3.32} = 2^{3-3.32} = 2^{-0.32}$

5. a) $4 \cdot 256 = 2^2 \cdot 2^8 = 2^{2+8} = 2^{10} = 1024$

b) $16 \cdot 128 = 2^4 \cdot 2^7 = 2^{4+7} = 2^{11} = 2048$

7. a) $6 \cdot 4 = 2^{2.585} \cdot 2^2 = 2^{2+2.585} = 2^{4.585} = 24$

b) $1 \cdot 6 = 2^0 \cdot 2^{2.585} = 2^{0+2.585} = 2^{2.585} = 6$

9. $2^{1.585} = 3$, $2^{2.585} = 6$, $2^{3.585} = 12$,

$2^{4.585} = 24$; each result is twice

the previous, each exponent is

one greater or 2^{n+1}.

11. a) $0.001 = 10^{-3}$

b) $\frac{1}{100000} = 10^{-5}$

c) $10,000 = 10^4$

d) $\frac{1}{100} = 10^{-2}$

e) $1,000,000 = 10^6$

f) $\frac{1}{1,000,000} = 10^{-6}$

13. a) $3000 = 3 \cdot 1000 = 3 \times 10^3$

b) $350 = 3.5 \cdot 100 = 3.5 \times 10^2$

c) $350000 = 3.5 \times 100,000 = 3.5 \times 10^5$

d) $0.00350 = 3.50 \div 1,000 = 3.50 \times 10^{-3}$

15. a) 1
b) 2
c) 2
d) 3

17. a) $29,979,000,000$ cm/sec

b) $0.000000000000000000000016726 g$

c) $-0.0000000000000000016022$ C

d) $-0.0000000000000000016022$ J

19. $(4.0 \cdot 10^{-2})(1.5 \cdot 10^6) = 4.0 \cdot 1.5 \cdot 10^{-2} \cdot 10^{-6}$

$= 6 \cdot 10^{-2+6} = 6 \cdot 10^4$

21. $\frac{2.4 \cdot 10^6}{0.3 \cdot 10^{-2}} = \frac{2.4}{0.3} \cdot 10^6 \cdot 10^2$

$= 8 \cdot 10^8$

Section 5.5

23. a) $(58 \text{ ft})^3 \cdot \dfrac{12 \text{ in}}{1 \text{ ft}} \cdot \dfrac{250,000 \text{ sheets}}{1 \text{ in}}$

$\approx 5.9 \times 10^{11} \text{ ft}^2$

b) $5.9 \times 10^{11} \text{ ft}^2 \cdot \left(\dfrac{1 \text{ mile}}{5280 \text{ ft}} \right)^2$

$\approx 2.1 \times 10^4 \text{ miles}^2$

c) New Hampshire and Vermont
combined = $18,966 \text{ mi}^2$
Tasmania = $26,200 \text{ mi}^2$
Costa Rica = $19,730 \text{ mi}^2$

25. a) $V = \sqrt{\dfrac{32.2 \frac{ft}{s^2}}{5.18 \times 10^{-4} \ ft^{-1}}}$

$= \sqrt{6.22 \times 10^4 \dfrac{ft^2}{s^2}}$

$V \approx 249 \dfrac{ft}{sec}$

b) $V = \sqrt{\dfrac{32.2 \frac{ft}{s^2}}{7.75 \times 10^{-4} \ ft^{-1}}}$

$= \sqrt{4.15 \times 10^4 \dfrac{ft^2}{s^2}}$

$V \approx 204 \dfrac{ft}{sec}$

c) $V = \sqrt{\dfrac{32.2 \frac{ft}{s^2}}{10.4 \times 10^{-4} \ ft^{-1}}}$

$= \sqrt{3.10 \times 10^4 \dfrac{ft^2}{s^2}}$

$V \approx 176 \dfrac{ft}{sec}$

27. a)

$V_{circ} = \sqrt{\dfrac{\left(6.67 \cdot 10^{-11} N \cdot m^2 \cdot kg^{-2} \right) \left(3.30 \cdot 10^{23} kg \right)}{3.24 \cdot 10^6 m}}$

$V_{circ} = \sqrt{6.79 \cdot 10^6 \dfrac{kg \cdot m}{sec^2} \cdot m \cdot kg^{-1}}$

$V_{circ} \approx 2.61 \cdot 10^3 \ m/sec$

b)

$V_{circ} = \sqrt{\dfrac{\left(6.67 \cdot 10^{-11} N \cdot m^2 \cdot kg^{-2} \right) \left(1.90 \cdot 10^{27} kg \right)}{8.14 \cdot 10^7 m}}$

$V_{circ} = \sqrt{1.56 \cdot 10^9 \dfrac{kg \cdot m}{sec^2} \cdot m \cdot kg^{-1}}$

$V_{circ} \approx 3.95 \cdot 10^4 \ m/sec$

c)

$V_{esc} = \sqrt{2} \cdot V_{circ}$

$V_{circ} = \sqrt{\dfrac{\left(6.67 \cdot 10^{-11} N \cdot m^2 \cdot kg^{-2} \right) \left(5.98 \cdot 10^{24} kg \right)}{7.18 \cdot 10^6 m}}$

$V_{circ} = \sqrt{5.55 \cdot 10^7 \dfrac{kg \cdot m}{sec^2} \cdot m \cdot kg^{-1}}$

$V_{circ} \approx 7.45 \cdot 10^3 \ m/sec$

$V_{esc} = \sqrt{2} \cdot 7.45 \cdot 10^3 \ m/sec$

$V_{esc} \approx 1.05 \cdot 10^4 \ m/sec$

d)

$V_{circ} = \sqrt{\dfrac{\left(6.67 \cdot 10^{-11} N \cdot m^2 \cdot kg^{-2} \right) \left(6.44 \cdot 10^{23} kg \right)}{4.19 \cdot 10^6 m}}$

$V_{circ} = \sqrt{1.03 \cdot 10^7 \dfrac{kg \cdot m}{sec^2} \cdot m \cdot kg^{-1}}$

$V_{circ} \approx 3.20 \cdot 10^3 \dfrac{m}{sec}$

$V_{esc} = \sqrt{2} \cdot 3.20 \cdot 10^3 \dfrac{m}{sec}$

$V_{esc} \approx 4.53 \cdot 10^3 \dfrac{m}{sec}$

29. T1-82 & T1-83; 9,999,999.

31. $10^{4.1} \approx 1.26 \cdot 10^4$

$10^{4.2} \approx 1.58 \cdot 10^4$

$10^{4.3} \approx 2.00 \cdot 10^4$

33. $10^{4.6} \approx 3.98 \cdot 10^4$

$10^{3.6} \approx 3.98 \cdot 10^3$

$10^{2.6} \approx 3.98 \cdot 10^2$

35. a) $10^1 \cdot 10^n = 10^{n+1}$

b) $2^1 \cdot 2^x = 2^{x+1}$

c) $3^1 \cdot 3^m = 3^{m+1}$

d) $2^1 \cdot 2^{2n} = 2^{2n+1}$

e) $2^3 \cdot 2^x = 2^{x+3}$

f) $9 \cdot 3^x = 3^2 \cdot 3^x = 3^{x+2}$

37. a) $t^{-1} = t_1^{-1} + t_2^{-1}$

$$\frac{1}{t} = \frac{1}{t_1} + \frac{1}{t_2}$$

b) $g = 9.81\, m \cdot sec^{-2}$

$$g = 9.81\, \frac{m}{sec^2}$$

c) $P = Ae^{-rt}$

$$P = \frac{a}{e^{rt}}$$

d) $d = 1\,g \cdot cm^{-3}$

$$d = \frac{1\,g}{cm^3}$$

e) $p = 31\,pounds \cdot in^{-2}$

$$p = \frac{31\,pounds}{in^2}$$

39. a) $186,000\, \dfrac{miles}{sec} \cdot \dfrac{3600\,sec}{hr}$

$\approx 6.70 \cdot 10^8\ mph$

b) Warp $2 = 2^3 \cdot 6.70 \cdot 10^8\ mph$

$\approx 5.36 \cdot 10^9\ mph$

c) Warp $3 = 3^3 \cdot 6.70 \cdot 10^8\ mph$

$\approx 1.81 \cdot 10^{10}\ mph$

d) Warp $14.1 = 14.1^3 \cdot 6.70 \cdot 10^8\ mph$

$\approx 1.88 \cdot 10^{12}\ mph$

Chapter 5 Review

1. $\dfrac{a^0 b^{-1}}{c^2} = \dfrac{1}{bc^2}$

3. $\left(\dfrac{2x}{y^2}\right)^4 = \dfrac{2^4 x^4}{\left(y^2\right)^4} = \dfrac{16x^4}{y^8}$

5. $\left(\dfrac{2x^2}{y}\right)^3 = \dfrac{2^3 (x^2)^3}{y^3} = \dfrac{8x^6}{y^3}$

7.

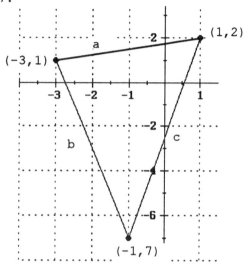

$(1,2)$
$(-3,1)$
$(-1,7)$

side a $d = \sqrt{(1-(-3))^2 + (2-1)^2}$
$\qquad d = \sqrt{16+1} = \sqrt{17}$

side b $d = \sqrt{(-1-(-3))^2 + (-7-1)^2}$
$\qquad d = \sqrt{4+64} = \sqrt{68}$
$\qquad d = \sqrt{4 \cdot 17} = 2\sqrt{17}$

side c $d = \sqrt{(-1-(-1))^2 + (2-(-7))^2}$
$\qquad d = \sqrt{4+81} = \sqrt{85}$
$a^2 + b^2 = c^2;$
$\left(\sqrt{17}\right)^2 + \left(2\sqrt{17}\right)^2 = \left(\sqrt{85}\right)^2$
$= 17 + 68 = 85 = c^2$
Right triangle

9. a) $\sqrt{9n} = 3\sqrt{n};\quad n \geq 0$

   b) $\sqrt{90n^2} = \sqrt{9n^2 \cdot 10} = 3|n|\sqrt{10}$

   c) $\sqrt{0.09x} = 0.3\sqrt{x};\quad x \geq 0$

   d) $\sqrt{900x^4} = 30x^2$

11. $\$1200\left(1 + \dfrac{0.06}{365}\right)^{365 \cdot 7} \approx \$1826.29$

13. $\$1200\left(1 + \dfrac{0.06}{365}\right)^{365 \cdot 2.5} \approx \$1394.18$

15. $\$2400 = \$1200\left(1 + \dfrac{r}{4}\right)^{4 \cdot 8}$

$2 = \left(1 + \dfrac{r}{4}\right)^{32}, \quad 2^{1/32} = 1 + \dfrac{r}{4}$

$2^{1/32} - 1 = \dfrac{r}{4}, \quad r = 4\left(2^{1/32} - 1\right)$

$r \approx 0.0876 = 8.76\%$

17. $\$8.00 = \$5.50(1 + r)^{10}$

$\left(\dfrac{8.00}{5.50}\right)^{1/10} - 1 = r$

$r \approx 0.0382 = 3.82\%$

19. $\$17,000 = \$7000(1 + r)^{12}$

$\left(\dfrac{17,000}{7000}\right)^{1/12} - 1 = r$

$r \approx 0.0767 = 7.67\%$

21. a) $125^{2/3} = \left(\sqrt[3]{125}\right)^2 = 5^2 = 25$

   b) $64^{1/3} = \sqrt[3]{64} = 4$

   c) $32^{2/5} = \left(\sqrt[5]{32}\right)^2 = 2^2 = 4$

23. a $\sqrt[3]{-8} = (-8)^{1/3} = -2$

b) $\sqrt[4]{\dfrac{625}{16}} = \left(\dfrac{625}{16}\right)^{1/4} = \dfrac{625^{1/4}}{16^{1/4}} = \dfrac{5}{2}$

25. a) $\left(b^4\right)^{3/4} = b^{4 \cdot 3/4} = b^3$

b) $\sqrt[3]{a^2} \cdot \sqrt[3]{a} = \sqrt[3]{a^3} = a$

c) $x^{3/4} \cdot x^{1/2} = x^{3/4 + 1/2} = x^{3/4 + 2/4} = x^{5/4}$

d) $\dfrac{b^{3/4}}{b^{1/2}} = b^{3/4 - 1/2} = b^{3/4 - 2/4} = b^{1/4}$

e) $b \cdot b^x = b^{x+1}$

27. a $\sqrt[3]{x} \cdot \sqrt[2]{x} = x^{1/3} \cdot x^{1/2} = x^{1/3 + 1/2}$
$= x^{2/6 + 3/6} = x^{5/6}; \quad x \geq 0$

b) $\sqrt{x^3} \cdot \sqrt{x^2} = x^{3/2} \cdot x^{2/2}$
$= x^{3/2 + 2/2} = x^{5/2}; \quad x \geq 0$

29. $F = \dfrac{Q_1 Q_2}{kr^2}$

$r^2 = \dfrac{Q_1 Q_2}{kF}$

$r = \sqrt{\dfrac{Q_1 Q_2}{kF}}$

31. $r^3 = a^2 p$

$r = \sqrt[3]{a^2 p}$

33. a) $x \approx \{0.2, 2\}$

b) From graph $x = 1$
$2 = \sqrt{5x - 1}$
$4 = 5x - 1$
$5x = 5$
$x = 1$

33. c) From graph $x \approx 0.5$
$\sqrt{8x} - 1 = 1$
$\sqrt{8x} = 2$
$8x = 4$
$x = \dfrac{4}{8} = \dfrac{1}{2}$

35. a) $x = 12$

b) From graph $x = 4$
$\sqrt{6x - 8} = 4$
$6x - 8 = 16$
$6x = 24$
$x = 4$

c) From graph $x = 0$
$\sqrt{3x} + 2 = 2$
$\sqrt{3x} = 0$
$x = 0$

37. a) $\sqrt{3} + 2\sqrt{12} = \sqrt{3} + 2\sqrt{4 \cdot 3}$
$= \sqrt{3} + 4\sqrt{3} = 5\sqrt{3}$

b) $\sqrt{9x} + \sqrt{x} = 3\sqrt{x} + \sqrt{x}$
$= 4\sqrt{x}; \; x \geq 0$

39. a) $\left(3 + \sqrt{6}\right)\left(3 - \sqrt{6}\right) = 3^2 - \left(\sqrt{6}\right)^2$
$= 9 - 6 = 3$

b) $\left(3 - \sqrt{x}\right)\left(3 - \sqrt{x}\right)$
$= 3^2 - 2 \cdot 3\left(\sqrt{x}\right) + \left(\sqrt{x}\right)^2 = 9 - 6\sqrt{x} + x$

41. $(8 - 2i)(8 + 2i) = 8^2 - (2i)^2 = 64 - 4(i^2)$
$= 64 - 4(-1) = 64 + 4 = 68$

43. a) $\dfrac{1}{b - \sqrt{a}} \cdot \dfrac{b + \sqrt{a}}{b + \sqrt{a}} = \dfrac{b + \sqrt{a}}{b^2 - a}$

b) $\dfrac{1}{2 + i} \cdot \dfrac{2 - i}{2 - i} = \dfrac{2 - i}{4 - i^2}$
$= \dfrac{2 - i}{4 - (-1)} = \dfrac{2 - i}{5}$

45. $\left(1 + i\sqrt{2}\right)^2 - 2\left(1 + i\sqrt{2}\right) + 3 \overset{?}{=} 0$

$1^2 + 2i\sqrt{2} + \left(i\sqrt{2}\right)^2 - 2 - 2i\sqrt{2} + 3 \overset{?}{=} 0$

$1 + 2i\sqrt{2} - 2 - 2 - 2i\sqrt{2} + 3 \overset{?}{=} 0$

$1 - 2 - 2 + 3 + 2i\sqrt{2} - 2i\sqrt{2} \overset{?}{=} 0$

$0 = 0$

$\left(1 - i\sqrt{2}\right)^2 - 2\left(1 - i\sqrt{2}\right) + 3 \overset{?}{=} 0$

$1^2 - 2i\sqrt{2} + \left(i\sqrt{2}\right)^2 - 2 + 2i\sqrt{2} + 3 \overset{?}{=} 0$

$1 - 2i\sqrt{2} - 2 - 2 + 2i\sqrt{2} + 3 \overset{?}{=} 0$

$0 = 0$

47. a) $4^x = 64 = 4^3$

$x = 3$

b) $2^x = 2 = 2^1$

$x = 1$

c) $a^x = \dfrac{1}{a} = a^{-1}$

$x = -1$

d) $b^n = 1 = b^0$

$n = 0$

49. a) $25^n = 125$

$5^{2n} = 5^3$

$2n = 3$

$n = \dfrac{3}{2}$

b) $27^n = \dfrac{1}{3}$

$3^{3n} = 3^{-1}$

$3n = -1$

$n = -\dfrac{1}{3}$

49. c) $\left(\dfrac{1}{25}\right)^n = 125$

$5^{-2n} = 5^3$

$-2n = 3$

$n = -\dfrac{3}{2}$

d) $\left(\dfrac{1}{16}\right)^n = 16$

$16^{-1n} = 16^1$

$-1n = 1$

$n = -1$

51. a) $\left(\dfrac{1}{100}\right)^n = 10$

$10^{-2n} = 10^1$

$-2n = 1$

$n = -\dfrac{1}{2}$

b) $(100)^n = 10$

$10^{2n} = 10^1$

$2n = 1$

$n = \dfrac{1}{2}$

c) $4^{x-4} = 64 = 4^3$

$x - 4 = 3$

$x = 7$

d) $27^{x+1} = 81$

$3^{3(x+1)} = 3^4$

$3x + 3 = 4$

$3x = 1$

$x = \dfrac{1}{3}$

Chapter 5 Review

53. a) $10^{2.2} = 1.585 \cdot 10^2$

b) $10^{1.2} = 1.585 \cdot 10^1$

c) $10^{0.2} = 1.585 \cdot 10^0$

d) $10^{3.2}$ Has the same decimal part.

55. a) $\sqrt[3]{8000} = \sqrt[3]{8 \cdot 10^3} = 2 \cdot 10 = 20$

b) $\sqrt[3]{800} = \sqrt[3]{8 \cdot 10^2} = 2\sqrt[3]{10^2} \approx 9.283$

c) $\sqrt[3]{80} = \sqrt[3]{8 \cdot 10^1} = 2\sqrt[3]{10^1} \approx 4.309$

d) $\sqrt[3]{8} = 2$

e) $\sqrt[3]{0.8} = \sqrt[3]{8 \cdot 10^{-1}}$
$= 2\sqrt[3]{10^{-1}} \approx 0.9283$

f) $\sqrt[3]{0.08} = \sqrt[3]{8 \cdot 10^{-2}}$
$= 2\sqrt[3]{10^{-2}} \approx 0.4309$

g) $\sqrt[3]{0.008} = \sqrt[3]{8 \cdot 10^{-3}}$
$= 2 \cdot 10^{-1} = 0.2$

57. $P_1 V_1^{1.3} = P_2 V_2^{1.3}$

$P_1 V_1^{13/10} = P_2 V_2^{13/10}$

$V_1^{13/10} = \dfrac{P_2 V_2^{13/10}}{P^1}$

$V_1 = \left(\dfrac{P_2 V_2^{13/10}}{P^1} \right)^{10/13} = \left(\dfrac{P_2}{P_1} \right)^{10/13} \cdot V_2$

$V_1 = V_2 \left(\dfrac{P_2}{P_1} \right)^{10/13}$

Chapter 5 Test

1. $\dfrac{a^0 b^2}{b^{-1}} = b^3$

2. $\dfrac{a^{-2} b^3}{(bc)^2} = \dfrac{b^3}{a^2 b^2 c^2} = \dfrac{b}{a^2 c^2}$

3. $\left(\dfrac{x^2 y}{3x}\right)^{-2} = \left(\dfrac{3x}{x^2 y}\right)^2 = \dfrac{3^2 x^2}{x^4 y^2} = \dfrac{9}{x^2 y^2}$

4. $(2x)^2 = 2^2 x^2 = 4x^2$

5. $\sqrt{4x^2} = \sqrt{4}\sqrt{x^2} = 2|x|$

6. $d_1 = \sqrt{\left(4 - (-2)\right)^2 + (1 - 2)^2}$

   $= \sqrt{36 + 1} = \sqrt{37}$

   $d_2 = \sqrt{(4 - 3)^2 + \left(1 - (-5)\right)^2}$

   $= \sqrt{1 + 36} = \sqrt{37}$

   $d_3 = \sqrt{\left(3 - (-2)\right)^2 + (-5 - 2)^2}$

   $= \sqrt{25 + 49} = \sqrt{74}$

   $d_1^2 + d_2^2 = 37 + 37 = 74 = d_3^2$
   Right isosceles triangle.

7. a) $\sqrt{36x} = 6\sqrt{x};\ x \geq 0$

   b) $\sqrt{0.36x^2} = 0.6|x|$

   c) $\sqrt{360x^4} = \sqrt{36 \cdot 10 \cdot x^4}$

   $= 6x^2 \sqrt{10}$

8. $a\left(\dfrac{b}{a}\right)^4 \overset{?}{=} b\left(\dfrac{b}{a}\right)^3$

   $\dfrac{a \cdot b^4}{a^4} \overset{?}{=} \dfrac{b \cdot b^3}{a^3}$

   $\dfrac{b^4}{a^3} = \dfrac{b^4}{a^3}$

9. $\$2000\left(1 + \dfrac{0.08}{4}\right)^{4 \cdot 3} = \$2536.48$

10. $\$18{,}000 = \$12{,}000(1 + r)^5$

    $\left(\dfrac{18{,}000}{12{,}000}\right)^{1/5} - 1 = r$

    $r \approx 0.0845 = 8.45\%$

11. $M = \dfrac{mg\ell}{\pi r^2 s},\ r^2 = \dfrac{mg\ell}{M\pi s}$

    $r = \sqrt{\dfrac{mg\ell}{\pi sM}}$

12. $F = \dfrac{mv^2}{r}\quad v^2 = \dfrac{Fr}{m}$

    $v = \sqrt{\dfrac{Fr}{m}}$

13. $64^{2/3} = \left(\sqrt[3]{64}\right)^2 = 4^2 = 16$

14. $32^{4/5} = \left(\sqrt[5]{32}\right)^4 = 2^4 = 16$

15. $81^{3/4} = \left(\sqrt[4]{81}\right)^3 = 3^3 = 27$

16. $\sqrt[4]{\dfrac{16}{81}} = \dfrac{\sqrt[4]{16}}{\sqrt[4]{81}} = \dfrac{2}{3}$

17. a) $\left(b^2\right)^{3/2} = b^{2 \cdot 3/2} = b^3$

    b) $b^{2/3} \cdot b^{1/3} = b^{2/3 + 1/3} = b^1 = b$

    c) $3 \cdot 3^n = 3^{n+1}$

18. a) $x = 8$

    b) $\sqrt{x + 1} = 2$

    $x + 1 = 4$

    $x = 3$

    c) $\sqrt{2x} - 1 = 1$

    $\sqrt{2x} = 2$

    $2x = 4$

    $x = 2$

19. a) $3^x = 27 = 3^3$

   $x = 3$

   b) $\left(\dfrac{1}{16}\right)^n = 64$

   $4^{-2n} = 4^3$

   $-2n = 3$

   $n = -\dfrac{3}{2}$

   c) $\dfrac{1}{1000} = 0.1^x$

   $10^{-3} = 10^{-x}$

   $-3 = -x$

   $x = 3$

   d) $4^{0.5x} = 16 = 4^2$

   $0.5x = 2$

   $x = 4$

20. a) $4^x = \dfrac{1}{64} = 4^{-3}$

   $x = -3$

   b) $100^n = 10$

   $10^{2n} = 10^1$

   $2n = 1$

   $n = \dfrac{1}{2}$

   c) $27^x = 9$

   $3^{3x} = 3^2$

   $3x = 2$

   $x = \dfrac{2}{3}$

   d) $4^{x+1} = 8$

   $2^{2(x+1)} = 2^3$

   $2x + 2 = 3$

   $2x = 1$

   $x = \dfrac{1}{2}$

21. a) $\left(2 - \sqrt{x}\right)\left(2 + \sqrt{x}\right) = 2^2 - \left(\sqrt{x}\right)^2 = 4 - x$

   b) $(8 - 3i)(3 + 8i) = 24 - 9i + 64i - 24i^2$

   $= 24 - 55i - 24(-1) = 48 - 55i$

   c) $\dfrac{1}{x + \sqrt{y}} \cdot \dfrac{x - \sqrt{y}}{x - \sqrt{y}} = \dfrac{x - \sqrt{y}}{x^2 - y}$

22. $(4 - 3i)^2 - 8(4 - 3i) + 25 \overset{?}{=} 0$

   $16 - 24i + 9i^2 - 32 + 24i + 25 \overset{?}{=} 0$

   $16 - 24i - 9 - 32 + 24i + 25 \overset{?}{=} 0$

   $16 - 9 - 32 + 25 - 24i + 24i \overset{?}{=} 0$

   $0 = 0$

23. a) $10^{2.1} = 125.89 = 1.26 \cdot 10^2$

   b) $10^{1.1} = 12.589 = 1.26 \cdot 10^1$

   c) $10^{0.1} = 1.2589 = 1.26 \cdot 10^0$
   The whole number part of the exponent determines the decimal point position. The 0.1 part of the exponent appears in each answer.

24. a) 0.000000000000000380

   b) 4,230,000,000,000,000

25. $3 \cdot 0.825$ billion $= 2.475$ billion
   (at least)
   2.475 billion $= 2,475,000,000$
   $2.475 \cdot 10^9$

Section 6.0

1. a) $\frac{1.5}{0.5} = 3,\quad \frac{4.5}{1.5} = 3,\quad \frac{13.5}{4.5} = 3$

   ratio, $r = 3$

   b) $a_n = a_1 \cdot r^{n-1},\quad a_1 = 0.5$

   $a_n = 0.5 \cdot 3^{n-1}$

   c) $y = 0.167 \cdot 3^x$

   d) $0.5 \cdot 3^{n-1} = 0.5 \cdot 3^n \cdot 3^{-1}$

   $= \frac{0.5 \cdot 3^n}{3} = \frac{0.5}{3} \cdot 3^n$

   $= 0.167 \cdot 3^n$

3. a) $\frac{20}{10} = 2,\quad \frac{40}{20} = 2,\quad \frac{80}{40} = 2$

   $r = 2$

   b) $a_1 = 10,\quad a_n = 10 \cdot 2^{n-1}$

   c) $y = 5 \cdot 2^x$

   d) $10 \cdot 2^{n-1} = 10 \cdot 2^n \cdot 2^{-1}$

   $= \frac{10 \cdot 2^n}{2} = \frac{10}{2} \cdot 2^n$

   $= 5 \cdot 2^n$

5. a) $\frac{1/2}{1/4} = 2,\quad \frac{1}{1/2} = 2,\quad \frac{2}{1} = 2,\quad \frac{4}{2} = 2$

   $r = 2$

   b) $a_1 = \frac{1}{4},\quad a_n = \frac{1}{4} \cdot 2^{n-1}$

   c) $y = 0.125 \cdot 2^x$

5. d) $\frac{1}{4} \cdot 2^{n-1} = \frac{1}{4} \cdot 2^n \cdot 2^{-1}$

   $= \frac{1}{4} \cdot \frac{2^n}{2} = \frac{1}{4} \cdot \frac{1}{2} \cdot 2^n$

   $= \frac{1}{8} \cdot 2^n = 0.125 \cdot 2^n$

7. a) $\frac{2}{1} = 2,\quad \frac{4}{2} = 2,\quad \frac{8}{4} = 2,\quad \frac{16}{8} = 2$

   $r = 2$

   b) $a_1 = 1,\quad a_n = 1 \cdot 2^{n-1}$

   c) $y = 0.5 \cdot 2^x$

   d) $1 \cdot 2^{n-1} = 1 \cdot 2^n \cdot 2^{-1}$

   $= 1 \cdot \frac{2^n}{2} = \frac{1}{2} \cdot 2^n = 0.5 \cdot 2^n$

9. a) $\frac{16}{32} = \frac{1}{2},\quad \frac{8}{16} = \frac{1}{2},\quad \frac{4}{8} = \frac{1}{2},\quad \frac{2}{4} = \frac{1}{2}$

   $r = \frac{1}{2}$

   b) $a_1 = 32,\quad a_n = 32 \cdot \left(\frac{1}{2}\right)^{n-1}$

   c) $y = 64 \cdot 0.5^x$

   d) $32 \cdot \left(\frac{1}{2}\right)^{n-1} = 32 \cdot \left(\frac{1}{2}\right)^n \cdot \left(\frac{1}{2}\right)^{-1}$

   $= 32 \cdot \left(\frac{1}{2}\right)^n \cdot 2 = 64 \cdot \left(\frac{1}{2}\right)^n = 64 \cdot 0.5^n$

11. a) $\frac{1}{3} = \frac{1}{3}$, $\frac{1/3}{1} = \frac{1}{3}$, $\frac{1/9}{1/3} = \frac{1}{3}$, $\frac{1/27}{1/9} = \frac{1}{3}$

      $r = \frac{1}{3}$

   b) $a_1 = 3$, $a_n = 3 \cdot \left(\frac{1}{3}\right)^{n-1}$

   c) $y = 9 \cdot 0.333^x$

   d) $3 \cdot \left(\frac{1}{3}\right)^{n-1} = 3 \cdot \left(\frac{1}{3}\right)^n \cdot \left(\frac{1}{3}\right)^{-1}$

     $= 3 \cdot \left(\frac{1}{3}\right)^n \cdot 3 = 9 \cdot \left(\frac{1}{3}\right)^n \approx 9 \cdot 0.333^n$

13. $2 \diagdown_{\diagup} 8 \diagdown_{\diagup} 18 \diagdown_{\diagup} 32$
      $6 \diagdown 10 \diagup 14$
        $4 \quad 4$

   Quadratic
   $2a = 4$, $a = 2$
   $3a + b = 6$, $3 \cdot 2 + b = 6$, $b = 0$
   $a + b + c = 2$, $2 + 0 + c = 2$, $c = 0$
   $y = 2x^2$

15. $0.5 \diagdown_{\diagup} 1.5 \diagdown_{\diagup} 4.5 \diagdown_{\diagup} 13.5$
     $1.0 \diagdown 3.0 \diagup 9.0$
       $2.0 \quad 6.0$

   $\frac{1.5}{0.5} = 3$, $\frac{4.5}{1.5} = 3$, $\frac{13.5}{4.5} = 3$

   Exponential
   $y = 0.5 \cdot 3^{x-1} \approx 0.167 \cdot 3^x$

17. $3 \diagdown_{\diagup} 7 \diagdown_{\diagup} 12 \diagdown_{\diagup} 18$
     $4 \diagdown_{\diagup} 5 \diagdown_{\diagup} 6$
       $1 \quad 1$

   Quadratic

   $2a = 1$, $a = \frac{1}{2}$
   $3a + b = 4$, $3\left(\frac{1}{2}\right) + b = 4$, $b = \frac{5}{2}$
   $a + b + c = 3$, $\frac{1}{2} + \frac{5}{2} + c = 3$, $c = 0$
   $y = \frac{1}{2}x^2 + \frac{5}{2}x = 0.5x^2 + 2.5x$

19. $30 \diagdown_{\diagup} 38 \diagdown_{\diagup} 46 \diagdown_{\diagup} 54 \diagdown_{\diagup} 62$
       $8 \quad\; 8 \quad\; 8 \quad\; 8$

   Linear

   $m = 8$, $b = 30 - 8 = 22$
   $y = 8x + 22$

21. $5 \diagdown_{\diagup} 10 \diagdown_{\diagup} 20 \diagdown_{\diagup} 40 \diagdown_{\diagup} 80$
     $5 \diagdown_{\diagup} 10 \diagdown_{\diagup} 20 \diagup 40$
       $5 \quad\;\; 10 \quad\; 20$

   Exponential
   $r = 2$
   $y = 5 \cdot 2^{x-1} = 2.5 \cdot 2^x$

23. $2^{x+2} = 2^x \cdot 2^2 = 4 \cdot 2^x$

25. $2^{x-1} = 2^x \cdot 2^{-1} = \frac{1}{2} \cdot 2^x = \frac{2^x}{2}$

27. $\left(\frac{1}{2}\right)^{-x} = \left(\left(\frac{1}{2}\right)^{-1}\right)^x = 2^x$

29. $\frac{2^n}{2} = 2^n \cdot 2^{-1} = 2^{n-1}$

31. $4 \cdot 2^n = 2^2 \cdot 2^n = 2^{n+2}$

33. $\frac{3^n}{3} = 3^n \cdot 3^{-1} = 3^{n-1}$

35. $9 \cdot 3^n = 3^2 \cdot 3^n = 3^{n+2}$

37. a) $\frac{1}{8} = 2^{-3}$, $2^{1+a} = 2^{-3}$
     $1 + a = -3$, $a = -4$
     $f(x) = 2^{x-4}$

   b) $4 = 2^2$, $2^{1+a} = 2^2$
     $1 + a = 2$, $a = 1$
     $f(x) = 2^{x+1}$

Section 6.0

39. a) $4 = \left(\dfrac{1}{2}\right)^{-2}$, $\left(\dfrac{1}{2}\right)^{1+a} = \left(\dfrac{1}{2}\right)^{-2}$

$1 + a = -2$, $a = -3$

$f(x) = \left(\dfrac{1}{2}\right)^{x-3}$

b) $8 = \left(\dfrac{1}{2}\right)^{-3}$, $\left(\dfrac{1}{2}\right)^{1+a} = \left(\dfrac{1}{2}\right)^{-3}$

$1 + a = -3$, $a = -4$

$f(x) = \left(\dfrac{1}{2}\right)^{x-4}$

41. $2, -6, 18, -54, 162$

$\dfrac{-6}{2} = -3$, $\dfrac{18}{-6} = -3$, $\dfrac{-54}{18} = -3$, $\dfrac{162}{-54} = -3$

$r = -3$

$a_n = 2 \cdot (-3)^{n-1}$

The base is negative so it is not an exponential function.

43. $\dfrac{-32}{64} = -\dfrac{1}{2}$, $\dfrac{16}{-32} = -\dfrac{1}{2}$, $\dfrac{-8}{16} = -\dfrac{1}{2}$, $\dfrac{4}{-8} = -\dfrac{1}{2}$

$r = -\dfrac{1}{2}$

$a_n = 64 \cdot \left(-\dfrac{1}{2}\right)^{n-1}$

Same as # 41.

45. $1 \diagdown_{0} 1 \diagdown_{1} 2 \diagdown_{1} 3 \diagdown_{2} 5 \diagdown_{3} 8$

Not geometric,
no common ratio.

47. $4 \diagdown_{1} 5 \diagdown_{4} 9 \diagdown_{5} 14 \diagdown_{9} 23 \diagdown_{14} 37$

Not geometric,
no common ratio.

49. a) $\dfrac{2''}{500} = 0.004$ in.

b) $1024 \cdot 0.004 \text{ in} = 4.096 \text{ in}$

c) 1 mile $= 5280$ ft $= 63360$ in

$\dfrac{63360}{0.004} = 15,840,000$ sheets

$2^x = 15,840,000$

$2^{23} = 8,388,608$, $2^{24} = 16,777,216$

24 folds

d) $93,000,000 \, (5280) \, (12) \approx 5.89 \times 10^{12}$ in

$2^{42} \approx 4.40 \times 10^{12}$, $2^{43} \approx 8.80 \times 10^{12}$

43 *folds*

1. a) a^x, base = a
   exponent = x
   exponential

   b) $4x^3$, base = x
   exponent = 3
   polynomial

   c) 2^{-x}, base = 2
   exponent = -x
   exponential

   d) x^a, base = x
   exponent = a
   polynomial

3. a) π^x, base = π
   exponent = x
   exponential

   b) $2x^e$, base = x
   exponent = e
   neither, exponent
   not an integer

   c) $a_1 r^{n-1}$, base = r
   exponent = n - 1
   exponential

   d) $100(1.06)^t$, base = 1.06
   exponent = t
   exponential

5. $3^{x-2} = 3^x \cdot 3^{-2} = 3^x \cdot \dfrac{1}{3^2} = \dfrac{3^x}{9}$

7. $2^{x-4} = 2^x \cdot 2^{-4} = 2^x \cdot \dfrac{1}{2^4} = \dfrac{2^x}{16}$

9. $\left(\dfrac{1}{4}\right)^{-x} = 4^x$

11. $16 \cdot 2^n = 2^4 \cdot 2^n = 2^{n+4}$

13. $\dfrac{1}{2} \cdot 2^n = 2^{-1} \cdot 2^n = 2^{n-1}$

15. $27 \cdot 3^n = 3^3 \cdot 3^n = 3^{n+3}$

17. $81\left(\dfrac{1}{3}\right)^{n-1} = 81 \cdot \left(\dfrac{1}{3}\right)^n \cdot \left(\dfrac{1}{3}\right)^{-1}$

   $= 81 \cdot 3 \cdot \left(\dfrac{1}{3}\right)^n = 243\left(\dfrac{1}{3}\right)^n$

19. 1, when x = 0, $5^x = 1$

21. $2b^0 = 2 \cdot 1 = 2$

23. $b^0 = 1$

25. a, in $y = ab^x$, represents
   the y – intercept value.

27. {}, graph is never below x-axis

29. {}, $y = 2^x$ is never negative

31. Answers vary, one is $\dfrac{1}{x^2}$

33. When x = 0, $b^x = 1$

35. Order of operations indicates
   that the exponent applies only
   to the letter b. $y = a \cdot b^x$.

37. Domain $\mathbb{R}$ for $y = 2^{-x}$
   Domain $x \geq 0$ for $y = x^{1/2}$
   $y = x^{1/2}$, starts at (0,0),
   increases slowly.
   $y = 2^{-x}$, starts large,
   approaches y = 0

39. $x^2 > 2^x$ for $x < -0.766$ & $2 < x < 4$
   for $x > 4$, 2^x rises faster.

41.

x	$y = 2^{x-1}$	$y = 2^x$	$y = 2^{x+1}$
-2	$2^{-3} = \frac{1}{8}$	$2^{-2} = \frac{1}{4}$	$2^{-1} = \frac{1}{2}$
-1	$2^{-2} = \frac{1}{4}$	$2^{-1} = \frac{1}{2}$	$2^0 = 1$
0	$2^{-1} = \frac{1}{2}$	$2^0 = 1$	$2^1 = 2$
1	$2^0 = 1$	$2^1 = 2$	$2^2 = 4$
2	$2^1 = 2$	$2^2 = 4$	$2^3 = 8$

Identify graphs by y-intercept:

a is $y = 2^{x+1}$

b is $y = 2^x$

c is $y = 2^{x-1}$

Adding shifts 1 unit to the left. Subtracting shifts 1 unit to the right.

43, 45, 47.

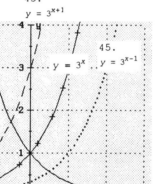

49. Graphs do not intersect.

$3^x \neq 3^{x+1}$ for all x.

51. a is $y = 10^x$

b is $y = 2.72^x$

c is $y = 2^x$

Identify graphs at x = 1.

53. a) Straight horizontal line,
   $y = 1$

b)

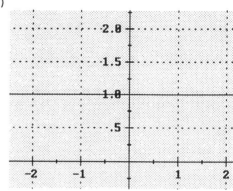

c) Horizontal line at y = 1.

d) Yes

e) $y = 1$

55. a)

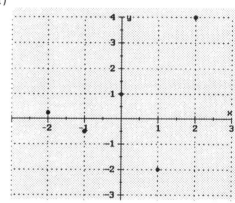

b) Points only

c) no

d) no

57. a) $x \approx 2.5$

b) $x \approx 5$

c) $y \approx 1.5 \text{ or } 2$

d) $y \approx 4$

e) $a \approx 1.3$

f) ≈ 2.5 units of time.

Section 6.1

59. $a_1 = 108, \ r = \dfrac{1}{2}$

$$a_n = 108\left(\dfrac{1}{2}\right)^{n-1}$$

$$y = 216\left(\dfrac{1}{2}\right)^n$$

$$216\left(\dfrac{1}{2}\right)^{10} \approx 0.211 \text{ inches}$$

61. $y = x$ is linear, $y = 0.909(1.1)^x$ is exponential, intersection $\approx (39.6, 39.6)$ is where option 2 becomes the better deal.

63. $\$1{,}000(1.02)^t = \$1{,}000{,}000$

$t \approx 349$ years

65. $\$1{,}000(1.10)^t = \$1{,}000{,}000$

$t \approx 73$ years

67. a) $13{,}758 = 5{,}036(1 + r)^{10}$

$$\dfrac{13{,}758}{5{,}036} = (1 + r)^{10}$$

$$\left(\dfrac{13{,}758}{5{,}036}\right)^{1/10} = 1 + r$$

$$r = \left(\dfrac{13{,}758}{5{,}036}\right)^{1/10} - 1$$

$r \approx 0.106$ or 10.6%

b) $A = 5036(1.106)^{60}$

$A \approx 2{,}100{,}000$

69. a) $39{,}089 = 40{,}108(1 + r)^{10}$

$$\dfrac{39{,}089}{40{,}108} = (1 + r)^{10}$$

$$\left(\dfrac{39{,}089}{40{,}108}\right)^{1/10} = 1 + r$$

$$r = \left(\dfrac{39{,}089}{40{,}108}\right)^{1/10} - 1$$

$r \approx -0.00257$ or $r \approx -0.26\%$

b) $A = 40{,}108\big(1 + (-0.0026)\big)^{60}$

$A = 40{,}108(0.9974)^{60}$

$A \approx 34{,}000$

1. a. $A_0 = 4$
   b. base = 3
   c. $k = 1$

3. a. $A_0 = 28,000$
   b. base = 0.97
   c. $k = \dfrac{1}{10}$

5. $y = 2^{x-1} = 2^x \cdot 2^{-1} = \dfrac{1}{2} \cdot 2^x$

7. $y = 3^{x+2} = 3^x \cdot 3^2 = 9 \cdot 3^x$

9. $y = 3^{x-3} = 3^x \cdot 3^{-3} = \dfrac{1}{27} \cdot 3^x$

11. $2^{kx} = \left(2^k\right)^x = (1.185)^x$

    $2^k = 1.185$

    $2^{0.24} \approx 1.181$

    $2^{0.25} \approx 1.189$

    $2^{0.245} \approx 1.185$

    $k \approx 0.245$

13. $2^{kx} = \left(2^k\right)^x = (1.06)^x$

    $2^k = 1.06$

    $2^{0.1} \approx 1.072$

    $2^{0.08} \approx 1.057$

    $2^{0.084} \approx 1.06$

    $k \approx 0.084$

15. a) $r^* = 0.08 \Rightarrow 8\%$
    b) $r^* = 0.06 \Rightarrow 6\%$

17. a) $r^* = -0.03 \Rightarrow -3\%$
    b) $0.95 = 1 - .05$

    $r^* = -0.05 \Rightarrow -5\%$

19. a) $2^{t/10} = \left(2^{1/10}\right)^t \approx 1.072^t$

    $r^* = 0.072 \Rightarrow 7.2\%$

    b) $2^{x/15} = \left(2^{1/15}\right)^x \approx 1.047^x$

    $r^* = 0.047 \Rightarrow 4.7\%$

21. a) $\left(\dfrac{1}{2}\right)^{2t} = \left(\left(\dfrac{1}{2}\right)^2\right)^t = \left(\dfrac{1}{4}\right)^t = \left(1 - \dfrac{3}{4}\right)^t$

    $r^* = -\dfrac{3}{4} = -.075 \Rightarrow -75\%$

    b) $\left(\dfrac{1}{3}\right)^{3x} = \left(\left(\dfrac{1}{3}\right)^3\right)^x = \left(\dfrac{1}{27}\right)^x = \left(1 - \dfrac{26}{27}\right)^x$

    $r^* = -\dfrac{26}{27} = -0.963 \Rightarrow -96.3\%$

23. $2^x = (1 + 1)^x,\ r^* = 1 \Rightarrow 100\%$

25. $\left(\dfrac{1}{2}\right)^x = \dfrac{1}{2^x} = 2^{-x}$,

    Functions are the same.

27.

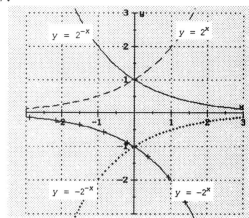

29.

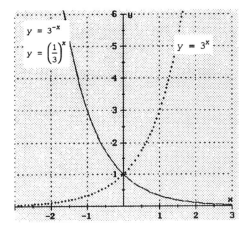

31. 1) y – intercept = 4, $4 \cdot 3^0 = 4$

3) y – intercept = 28,000,

$28,000(0.97)^{0/10}$

5) y – intercept = $\dfrac{1}{2}$,

$2^{0-1} = 2^{-1} = \dfrac{1}{2}$

33. $(-9)^{1/2} = \sqrt{-9}$

35.

$\dfrac{A_0}{2} = A_0 b^{kt}$ now divide both sides

by $A_0 \Rightarrow \dfrac{1}{2} = b^{kt}$

37. As interest rate increases doubling time decreases; inverse variation.

39. a) $72 \div 17 = 4.2\%$

$2 = 2^{k \cdot 17}$, $17k = 1$

$k = \dfrac{1}{17}$

$\left(2^{1/17}\right)^t = (1.042)^t$

$r^* = 0.042 \Rightarrow 4.2\%$

41. a) $72 \div 35 = 2.1\%$

b) $2 = 2^{k \cdot 35}$, $k \cdot 35 = 1$, $k = \dfrac{1}{35}$

$\left(2^{1/35}\right)^t = (1.020)^t$

$r^* = 0.020 \Rightarrow 2.0\%$

43. a) $72 \div 100 = 0.72\%$

b) $2 = 2^{k \cdot 100}$, $k \cdot 100 = 1$, $k = \dfrac{1}{100}$

$\left(2^{1/100}\right)^t = (1.007)^t$

$r^* = 0.007 \Rightarrow 0.7\%$

45. Using ExpReg on calculator

$y = 24780(1.049)^x$

$r^* = 0.049 \Rightarrow 4.9\%$

year 2000, $y \approx 705,000$

$y = 24780(1.049)^{70} \approx 705,000$

(note: if student uses a & b

from ExpReg $y = a \cdot b^x$ to

predict population ans. is

$y = 717,000$ due to round off

error.)

47. $y \approx 42738(0.998)^x$

$(0.998)^x = (1 - 0.002)^x$

$r^* = -0.002 \Rightarrow -0.2\%$

year 2000 $y \approx 37,000$

$y = 42738(0.998)^{70} \approx 37,000$

49. a) $y \approx 25600(0.87)^x$

b) $y \approx 50(0.87)^x$ The base (0.87),

remains the same but the

initial value (for t = 0)

changes.

Section 6.2

51. $y \approx 49.28(0.626)^x$

    impact year 2000 $\approx$ \$0.285 million

    $y = 49.28(0.626)^{11} = 0.285$

53. a) $1\,yr = 365\,days$

    $365\,days \cdot 24\,\dfrac{hrs}{day} = 8760\,hrs$

    $\dfrac{8760\,hrs}{26.3\,\dfrac{hrs}{half\,life}} \approx 333\,half\,lives$

    b) $y = 2{,}500{,}000(0.974)^x$

55. a) $\dfrac{365\,days}{14.3\,\dfrac{days}{half\,life}} \approx 26\,half\,lives$

    b) $y = 230{,}000(0.953)^x$

57. a) $\dfrac{8760\,hrs}{15\,\dfrac{hrs}{half\,life}} \approx 584\,half\,lives$

    b) $y = 12{,}000{,}000(0.955)^x$

1. a $400 = 4 \cdot 100 = 4 \cdot 10^2$

   $10^n = 400 = 4 \cdot 10^2$

   $4 = 10^{0.60206}$

   $10^n = 10^{0.60206} \cdot 10^2$

   $10^n = 10^{2.60206}$

   $n = 2.60206$

   b) $5000 = 5 \cdot 1000 = 5 \cdot 10^3$

   $10^n = 5000 = 5 \cdot 10^3$

   $\left(5 = 10^{0.69897}\right)$

   $10^n = 10^{0.69897} \cdot 10^3$

   $10^n = 10^{3.69897}$

   $n = 3.69897$

   c) $5,000,000 = 5 \cdot 1,000,000 = 5 \cdot 10^6$

   $10^n = 5,000,000 = 5 \cdot 10^6$

   $5 = 10^{0.69897}$

   $10^n = 10^{0.69897} \cdot 10^6$

   $10^n = 10^{6.69897}$

   $n = 6.69897$

   d) $500,000 = 5 \cdot 100,000 = 5 \cdot 10^5$

   $10^n = 500,000 = 5 \cdot 10^5$

   $\left(5 = 10^{0.69897}\right)$

   $10^n = 10^{0.69897} \cdot 10^5$

   $10^n = 10^{5.69897}$

   $n = 5.69897$

3. a) 0

   b) 0.77815

   c) 1

   d) 1.77815

   e) 2

   f) 2.77815

5. a) -2

   b) -1

   c) -0.22185

   d) -1.22185

   e) -2.22185

   f) -5.22185

7. a) Decimal portion is the same. All are positive.

   b) Decimal portion is the same. All are negative.

9. a) base $= 3$

   b) base $= 10$

   c) base $= 4$

   d) base $= 5$

11. a) $10^3 = 1000$

   b) $10^0 = 1$

   c) $3^4 = 81$

   d) $10^2 = 100$

13. a) $10^{-3} = 0.001$

   b) $4^0 = 1$

   c) $m^k = n$

   d) $5^{-2} = \dfrac{1}{25}$

15. a) $\log_2 32 = 5$

   b) $\log_2 2 = 1$

   c) $\log_2 1 = 0$

   d) $\log 100 = 2$

17. a) $\log 0.001 = -3$

   b) $\log_f g = d$

   c) $\log_3\left(\dfrac{1}{9}\right) = -2$

   d) $\log_4 1 = 0$

19. a) $7^2 = x$, $x = 49$

b) $3^x = 3$, $x = 1$

c) $10^x = 0.01$, $10^x = 10^{-2}$, $x = -2$

d) $10^0 = x$, $x = 1$

e) $x^3 = 64$, $x^3 = 4^3$, $x = 4$

21. a) $2^8 = x$, $x = 256$

b) $10^{-2} = x$, $x = 0.01$

c) $2^x = \dfrac{1}{4}$, $2^x = 2^{-2}$, $x = -2$

d) $10^x = 1$, $10^x = 10^0$, $x = 0$

e) $x^2 = 100$, $x^2 = 10^2$, $x = 10$

23. a) $\log 17 = x$, $x \approx 1.23045$

b) $\log 125 = x$, $x \approx 2.09691$

c) $\log x = 2.5$, $x \approx 316.228$

d) $\log 0.05 = x$, $x \approx -1.30103$

25. a) $10^{3.29907} \approx 1991$

b) $10^{0.9132} \approx 8.188$

c) $10^{2.5} \approx 316.2278$

d) $10^{1.69897} = x$, $x \approx 50$

27. a) $10^{1.30103} = x$, $x \approx 20$

b) $10^{2.30103} = x$, $x \approx 200$

c) $10^{-3} = 0.001$

d) $10^1 = 10$

29. Antitrust: opposed to large combinations of corporations.

31. Antibody: a substance that counters foreign organisms in the body.

33. a) $\log_3 10 = x$

$\dfrac{\log 10}{\log 3} \approx \dfrac{1}{0.4771}$

$x \approx 2.0959$

b) $\dfrac{\log 6}{\log 4} \approx 1.29248 = x$

c) $2 = 1.07^x$

$\log_{1.07} 2 = x$

$\dfrac{\log 2}{\log 1.07} \approx 10.24477 = x$

d) $\dfrac{1}{2} = 0.95^x$

$\log_{0.95}\left(\dfrac{1}{2}\right) = x$

$\dfrac{\log\left(\dfrac{1}{2}\right)}{\log 0.95} \approx 13.5134 = x$

e) $\dfrac{1}{2} = 0.5^x = \left(\dfrac{1}{2}\right)^x$

$x = 1$

35. $5^0 = x$, $\log_5 x = 0$, $x = 1$

$x^3 = 64$, $\log_x 64 = 3$, $x = 4$

$10^x = -10$, $\log(-10) = x$, $\{\ \}$

37. $2 = 1.08^t$

$t = \log_{1.08} 2 = \dfrac{\log 2}{\log 1.08} \approx 9$ years

39. $20 = 0.909(1.1)^x$

$\dfrac{20}{0.909} = 1.1^x$

$x = \log_{1.1}\left(\dfrac{20}{0.909}\right) = \dfrac{\log\left(\dfrac{20}{0.909}\right)}{\log 1.1} \approx 33$ yrs

Section 6.3

41. $h = 144\left(\dfrac{3}{4}\right)^{n-1}$

a) $60.75 = 144\left(\dfrac{3}{4}\right)^{n-1}$

$\dfrac{60.75}{144} = \left(\dfrac{3}{4}\right)^{n} \cdot \left(\dfrac{3}{4}\right)^{-1} = \left(\dfrac{3}{4}\right)^{n} \cdot \dfrac{4}{3}$

$\dfrac{3}{4} \cdot \dfrac{60.75}{144} = \left(\dfrac{3}{4}\right)^{n} = 0.75^{n}$

$n = \log_{0.75}\left(\dfrac{3 \cdot 60.75}{4 \cdot 144}\right) = \dfrac{\log\left(\dfrac{3 \cdot 60.75}{4 \cdot 144}\right)}{\log(0.75)}$

$n \approx 4$

b) $20 = 144\left(\dfrac{3}{4}\right)^{n-1}$

$\dfrac{20}{144} = \left(\dfrac{3}{4}\right)^{n} \cdot \left(\dfrac{4}{3}\right)$

$\dfrac{3}{4} \cdot \dfrac{20}{144} = 0.75^{n}$

$n = \log_{0.75}\left(\dfrac{3 \cdot 20}{4 \cdot 144}\right) = \dfrac{\log\left(\dfrac{60}{576}\right)}{\log(0.75)}$

$n \approx 7.86$, when h = 20

for h < 20 n ≥ 8

1. a. $\dfrac{9}{3} = 3$, $\dfrac{27}{9} = 3$, $\dfrac{81}{27} = 3$, $\dfrac{243}{81} = 3$

   Geometric, $r = 3$

   $a_n = 3 \cdot 3^{n-1}$, $f(x) = 3^x$

   $3 \cdot 3^{n-1} = 3 \cdot 3^n \cdot 3^{-1} = 3 \cdot \dfrac{1}{3} \cdot 3^n = 3^n$

   b) $-1\diagdown\diagup 5\diagdown\diagup 11\diagdown\diagup 17\diagdown\diagup 23$
   $\qquad\quad 6 \quad\; 6 \quad\; 6 \quad\;\; 6$

   Linear, $m = 6$,

   $-1 = 6(1) + b$, $b = -7$
   $y = 6x - 7$

   c) $\dfrac{6}{3} = 2$, $\dfrac{12}{6} = 2$, $\dfrac{24}{12} = 2$, $\dfrac{48}{24} = 2$

   Geometric, $r = 2$

   $a_n = 3 \cdot 2^{n-1}$, $f(x) = 1.5 \cdot 2^x$

   $3 \cdot 2^{n-1} = 3 \cdot 2^n \cdot 2^{-1}$

   $\qquad\qquad = 3 \cdot 2^n \cdot \dfrac{1}{2} = 1.5 \cdot 2^n$

   d) $\dfrac{27}{81} = \dfrac{1}{3}$, $\dfrac{9}{27} = \dfrac{1}{3}$, $\dfrac{3}{9} = \dfrac{1}{3}$

   Geometric, $r = \dfrac{1}{3}$

   $a_n = 81 \cdot \left(\dfrac{1}{3}\right)^{n-1}$, $f(x) = 243 \cdot \left(\dfrac{1}{3}\right)^x$

   $81 \cdot \left(\dfrac{1}{3}\right)^{n-1} = 81 \cdot \left(\dfrac{1}{3}\right)^n \cdot \left(\dfrac{1}{3}\right)^{-1}$

   $\qquad\qquad = 81 \cdot \left(\dfrac{1}{3}\right)^n \cdot 3 = 243 \cdot \left(\dfrac{1}{3}\right)^n$

   e) $1\diagdown\diagup 6\diagdown\diagup 15\diagdown\diagup 28\diagdown\diagup 45$
   $\quad\; 5\diagdown\; 9\diagdown\;\; 13\diagdown\;\; 17$
   $\qquad\; 4 \quad\;\; 4 \quad\;\; 4$

   Quadratic
   $2a = 4$, $a = 2$
   $3(2) + b = 5$, $b = -1$
   $2 + (-1) + c = 1$, $c = 0$
   $y = 2x^2 - x$

1. f) $\dfrac{1/4}{1/8} = 2$, $\dfrac{1/2}{1/4} = 2$, $\dfrac{1}{1/2} = 2$ $\dfrac{2}{1} = 2$

   Geometric, $r = 2$

   $a_n = \left(\dfrac{1}{8}\right) \cdot 2^{n-1}$, $f(x) = 0.0625 \cdot 2^x$

   $\left(\dfrac{1}{8}\right) \cdot 2^{n-1} = \left(\dfrac{1}{8}\right) \cdot 2^n \cdot 2^{-1}$

   $= \left(\dfrac{1}{8}\right) \cdot 2^n \cdot \left(\dfrac{1}{2}\right) = \left(\dfrac{1}{8}\right) \cdot \left(\dfrac{1}{2}\right) \cdot 2^n$

   $= \dfrac{1}{16} \cdot 2^n = 0.0625 \cdot 2^n$

2.

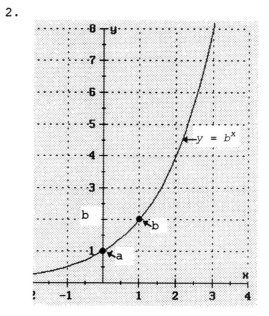

3. a) Graphs start near the x-axis on the left, rise rapidly on right side of y-axis.

   b) $y = 2^{x-1}$ is $y = 2^x$ scaled by a factor of $\dfrac{1}{2}$. $\left(y = \dfrac{1}{2} \cdot 2^x\right)$

   $y = 2^{x-1}$ is 2^x shifted right one unit in x.

   c) They never intersect.
   $2^x \neq 2^{x-1}$

Mid-Chapter Six Test

4. a)

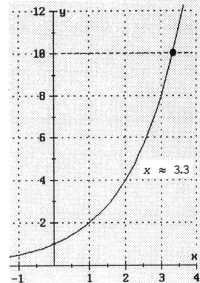

$x \approx 3.3$

b) $2^x = 10$ or $\log_2 10 = x$

5. a) $f(x) \approx 1.99 \cdot 1.046^x$

b) year 2000 $\approx \$3.75$

6. a) $2^x = 32,\quad 2^x = 2^5$

b) $\left(\frac{1}{2}\right)^x = 4,\quad \left(\frac{1}{2}\right)^x = 2^2$

$\left(\frac{1}{2}\right)^x = \left(\frac{1}{2}\right)^{-2},\quad x = -2$

c) $\left(\frac{1}{2}\right)^x = \frac{1}{16},\quad \left(\frac{1}{2}\right)^x = \left(\frac{1}{2}\right)^4$

$x = 4$

d) $5^x = \frac{1}{125},\quad 5^x = \left(\frac{1}{5}\right)^3$

$5^x = 5^{-3},\quad x = -3$

e) $10^x = -10,\quad \{\,\}$

f) $10^x = 0.01,\quad 10^x = 10^{-2}$

$x = -2$

7. a) $\log_4 x = -1,\quad 4^{-1} = x$

$x = \frac{1}{4}$

b) $\log 0.0001 = x,\quad 10^x = 0.0001$

$10^x = 10^{-4},\quad x = -4$

c) $\log_3 x = 3,\quad 3^3 = x$

$x = 27$

d) $\log_2 x = 0,\quad 2^0 = x$

$x = 1$

e) $\log_4 2 = x,\quad 4^x = 2$

$\left(2^2\right)^x = 2,\quad 2^{2x} = 2^1$

$2x = 1,\quad x = \frac{1}{2}$

f) $\log_4 -2 = x,\quad 4^x = -2,\quad \{\,\}$

8. $y = \frac{1}{2} \cdot 3^x$ grows rapidly
   as x gets large.

   $y = 2 \cdot 3^{-x}$ approaches zero
   as x gets large.

   a) Constants 1/2 and 2 are
      y-intercepts

   b) The negative exponent
      indicates decay.

9.
   $10^x = 15,\ \log 15 = x,\ x \approx 1.17609$

   $10^x = 13,\ \log 13 = x,\ x \approx 1.11394$

   $3^x = 24,\ \log_3 24 = x,\ x = \frac{\log 24}{\log 3} \approx 2.89279$

   $3^9 = x,\ \log_3 x = 9,\ x = 19683$

   $4^{x+1} = 50,\ \log_4 50 = x + 1,$

   $x = \frac{\log 50}{\log 4} - 1 \approx 1.82193$

1. a) $x \approx 2.8$ Second fig. is easier to read for smaller values of x.

   b) $2^x = 7$, $\log_2 7 = x$

   $$x = \frac{\log 7}{\log 2} \approx 2.8074$$

3. Graph becomes a straight line.

5. 2 locations, log2 and log20. Answers will vary, possible answers are 2,20,200.

7.

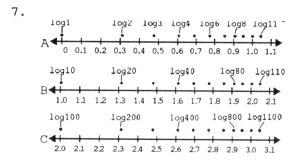

   d) Dots are in the same relative positions because $\log 20 = 1 + \log 2$, etc.

9. $\approx 22 ft$

11. 60 to 70 yrs

13. yrs 2 to 6, wind breakage.

15. A straight line.

17. a) $y \approx 52.2(0.95)^x$

   b) The deeper the water the less amount of light penetration. Light penetration decays exponentially with the depth of the water.

   c) $\log_{10} y = -0.02x + 1.72$

   d) $y = 10^{(-0.02x+1.72)}$

   $$y = 10^{-0.02x} \cdot 10^{1.72}$$

   $$y = \left(10^{-0.02}\right)^x \cdot 10^{1.72}$$

   $$y = (0.95)^x \cdot 52.5 = 52.5(0.95)^x$$

19. a) $y = 801(0.868)^x$

   b) Pressure decreases as altitude increases. Pressure decays exponentially with altitude.

   c) $\log_{10} y = -0.0617x + 2.904$

   d) $y = 10^{(-0.0617x+2.904)}$

   $$y = 10^{-0.0617} \cdot 10^{2.904}$$

   $$y = \left(10^{-0.0617}\right)^x \cdot 10^{2.904}$$

   $$y = 802(0.868)^x$$

21.

$$y = \log_2 0.5 = \frac{\log 0.5}{\log 2} = -1 \quad (0.5, -1)$$

$$y = \log_2 1 = \frac{\log 1}{\log 2} = 0 \quad (1, 0)$$

$$y = \log_2 2 = \frac{\log 2}{\log 2} = 1 \quad (2, 1)$$

$$y = \log_2 4 = \frac{\log 4}{\log 2} = 2 \quad (4, 2)$$

$$y = \log_2 8 = \frac{\log 8}{\log 2} = 3, \quad (8, 3)$$

23. $(1,0)$; $f(2) = 1$

25. False; $y = \log_b x = \dfrac{\log x}{\log b}$

    when $b > 10$, $\log b > 1$

    $\dfrac{\log x}{\log b} < \log x$

27. Graphs are symmetric over $y = x$;

    when $x = 0$, $2^x = 1$; $\log_2 x = 0$ when $x = 1$

    when $x = 1$, $2^x = 2$; $\log_2 x = 1$ when $x = 2$

    $(2, 4)$ is on y_2 and $(4, 2)$ is on y_3

29. a) $-\log\left(1.0 \cdot 10^{-14}\right) = 14$ base

    b) $-\log(0.2) \approx 0.7$ acid

    c) $-\log\left(5.01 \cdot 10^{-3}\right) \approx 2.3$ acid

    d) $-\log\left(1 \cdot 10^{-10}\right) = 10$ base

    e) $-\log\left(1.26 \cdot 10^{-3}\right) = 2.9$ acid

31. See # 29 above.

33. a) $4.1 = -\log H^+$

    $10^{-4.1} = H^+ = 7.94 \cdot 10^{-5}$ M

    b) $-0.5 = -\log H^+$

    $10^{0.5} = H^+ = 3.16$ M

    c) $12.0 = -\log H^+$

    $10^{-12.0} = H^+ = 1 \cdot 10^{-12}$ M

    d) $3.5 = -\log H^+$

    $10^{-3.5} = H^+ = 3.16 \cdot 10^{-4}$ M

35. pH is negative when $[H^+] > 1$;
    pH is positive when $[H^+] < 1$.

37. a) $R = \log 31{,}623{,}000 = 7.5$

    b) $R = \log 125{,}890{,}000 = 8.1$

39. a) $8.6 = \log I$

    $10^{8.6} = I \approx 398{,}107{,}000$

    b) $6.8 = \log I$

    $10^{6.8} = I \approx 6{,}310{,}000$

41. $I_{89} = 10^{6.9} \approx 7{,}943{,}000$

    $I_{94} = 10^{6.8} \approx 6{,}310{,}000$

    $\dfrac{I_{89}}{I_{94}} = \dfrac{7{,}943{,}000}{6{,}310{,}000} \approx 1.3$

43. I of 7.2 quake $= 10^{7.2} \approx 15{,}849{,}000$

    I of 6.8 quake $\approx 6{,}310{,}000$

    $4 \cdot 6{,}310{,}000 \approx 25{,}240{,}000$

    Announcer was incorrect, he may have subtracted 6.8 from 7.2 and multiplied by 10.

Section 6.5

1. a) $\ln y = 1$, $\log_e y = 1$, $y = e^1$, $y = e$

   b) $\ln y = 0$, $\log_e y = 0$, $y = e^0$, $y = 1$

   c) $y = \ln(-1)$, $y = \log_e(-1)$, $e^y = -1$, {}

   d) $y = \ln e^2$, $y = \log_e e^2$, $e^y = e^2$, $y = 2$

   e) $y = \ln e^e$, $y = \log_e e^e$, $e^y = e^e$, $y = e$

3. a) $e^2 \approx 7.389$

   b) $e^\pi \approx 23.141$

   c) $2\sqrt{e} \approx 3.297$

5. $e^{\wedge}(e^{\wedge}1)$

7. $x = 0$; Every number, except 0, raised to to zero power $= 1$.

9. a) $e^x = 4$, $\ln 4 = x$, $x \approx 1.3863$

   b) $\ln x = -2$, $e^{-2} = x$, $x \approx 0.1353$

11. a) $\ln x = 1.5$, $x = e^{1.5}$, $x \approx 4.4817$

    b) $e^x = 2$, $x = \ln 2$, $x \approx 0.6931$

13. a) $A = \$1000\left(1 + \dfrac{0.08}{12}\right)^{1 \cdot 12}$

    $A = \$1083.00$

    b) $A = \$1000 \cdot e^{0.08 \cdot 1}$

    $A = \$1083.29$

15. a) $A = \$1000\left(1 + \dfrac{0.10}{12}\right)^{1 \cdot 12}$

    $A = \$1104.71$

    b) $A = \$1000 \cdot e^{0.10 \cdot 1}$

    $A = \$1105.17$

17. a) $A = \$1000\left(1 + \dfrac{0.045}{12}\right)^{1 \cdot 12}$

    $A = \$1045.94$

    b) $A = \$1000 \cdot e^{0.045 \cdot 1}$

    $A = \$1046.03$

19. $3P = Pe^{0.08t}$

    $3 = e^{0.08t}$

    $\ln 3 = 0.08t$

    $\dfrac{\ln 3}{0.08} = t$

    $t = \dfrac{\ln 3}{0.08} \approx 13.7$ yrs

21. $3P = Pe^{rt}$

    $3 = e^{rt}$

    $\ln 3 = rt$

    $t = \dfrac{\ln 3}{r}$

23. $\dfrac{P}{2} = Pe^{-0.08t}$

    $\dfrac{1}{2} = e^{-0.08t}$

    $\ln\left(\dfrac{1}{2}\right) = -0.08t$

    $t = \dfrac{\ln\left(\dfrac{1}{2}\right)}{-0.08} \approx 8.7$ yrs

25. Double

27. a) $\dfrac{\ln 8}{\ln 6} \approx 1.1606$

    $\dfrac{\log 8}{\log 6} \approx 1.1606$

    b) $\log 8 \approx 0.903$ $\ln 8 \approx 2.079$

    $\log 6 \approx 0.778$ $\ln 6 \approx 1.792$

    c) Ratios are equal but logarithms are not.

29. $e^{-rt} = \dfrac{1}{e^{rt}}$

31. a) $6! = 6 \cdot 5 \cdot 4 \cdot 3 \cdot 2 \cdot 1 = 720$

    b) $9! = 9 \cdot 8 \cdot 7 \cdot 6 \cdot 5 \cdot 4 \cdot 3 \cdot 2 \cdot 1$

    $= 362880$

33.

$$1 - \frac{1}{2} + \frac{1}{3} - \frac{1}{4} + \frac{1}{5} - \frac{1}{6} + \frac{1}{7} - \frac{1}{8} + \frac{1}{9}$$

$$- \frac{1}{10} + \frac{1}{11} - \frac{1}{12} + \frac{1}{13} - \frac{1}{14} + \frac{1}{15} \approx 0.72537$$

$$\ln 2 \approx 0.69315$$

35. a)

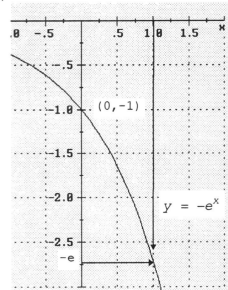

b)

37. a)

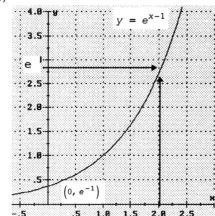

b)

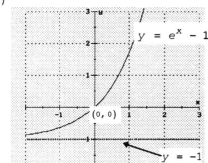

39. a)

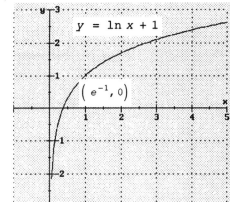

$$y = \ln x + 1$$

$\left(e^{-1}, 0\right)$

b)

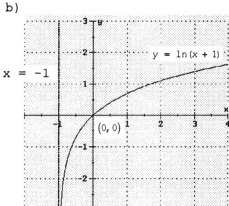

$x = -1$

$y = \ln(x + 1)$

$(0, 0)$

41.

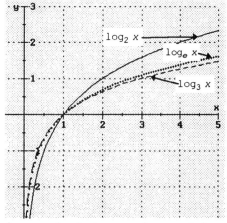

$\log_2 x$

$\log_e x$

$\log_3 x$

$\log_e x$ is between $\log_2 x$ & $\log_3 x$ and is nearer to $\log_3 x$.

43. a) $f(0) = 1849568e^{-0.0147 \cdot 0} = 1{,}849{,}568$

b) $f(20) = 1849568e^{-0.0147 \cdot 20} = 1{,}380{,}000$

c) $f(40) = 1849568e^{-0.0147 \cdot 40} = 1{,}030{,}000$

d) The sign of the exponent indicates increasing or decreasing. Negative is decreasing.

e) $\frac{1}{2}(1849568) = 1849568e^{-0.0147 \cdot t}$

$\frac{1}{2} = e^{-0.0147 \cdot t}, \quad \ln\left(\frac{1}{2}\right) = -0.0147t$

$t = \dfrac{\ln\left(\dfrac{1}{2}\right)}{-0.0147} \approx 47 \; years$

$1950 + 47 = 1997$

f) $f(50) = 1849568e^{-0.0147 \cdot 50} \approx 887{,}000$

45. $\ln k = \dfrac{-Ea}{RT} + C$

$k = e^{\left(\frac{-Ea}{RT} + C\right)}$

47. $E = \dfrac{-RT}{nF} \ln[H^+]$

$\dfrac{EnF}{-RT} = \ln[H^+]$

$[H^+] = e^{\left(-\frac{EnF}{RT}\right)}$

49.

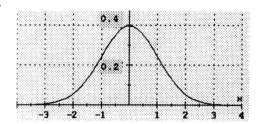

51. $y = \dfrac{1}{\sqrt{2\pi}} e^{-\frac{0^2}{2}} = \dfrac{1}{\sqrt{2\pi}}$

$y \approx 0.39894$

Section 6.5

53. Graph gives:

Terminal velocity $\approx 176\,\dfrac{ft}{sec}$

$t \approx 28\,sec$
To check:

$v = \dfrac{32.2}{0.183}\left(1 - e^{-0.183\cdot 28}\right)$

$v \approx 175\,\dfrac{ft}{sec}\quad \left(within\ 1\,\dfrac{ft}{sec}\right)$

55. Graph gives:

Terminal velocity $\approx 203\,\dfrac{ft}{sec}$

$t \approx 33\,sec$
To check:

$v = \dfrac{32.2}{0.1586}\left(1 - e^{-0.1586\cdot 33}\right)$

$v \approx 202\,\dfrac{ft}{sec}\quad \left(within\ 1\,\dfrac{ft}{sec}\right)$

57. $\left(\dfrac{1}{2}\right)2{,}500{,}000 = 2{,}500{,}000e^{k\cdot 26.3}$

$\left(\dfrac{1}{2}\right) = e^{26.3k},\quad \ln\left(\dfrac{1}{2}\right) = 26.3k$

$k = \dfrac{\ln\left(\dfrac{1}{2}\right)}{26.3} \approx -0.0264$

59. $\left(\dfrac{1}{2}\right)230{,}000 = 230{,}000e^{k\cdot 14.3}$

$\left(\dfrac{1}{2}\right) = e^{14.3k},\quad \ln\left(\dfrac{1}{2}\right) = 14.3k$

$k = \dfrac{\ln\left(\dfrac{1}{2}\right)}{14.3} \approx -0.0485$

61. No.

63. As x gets large e^{-bx} gets small.
Model looks like $y = c/(1 + 0)$
approaches $y = c$

65. The curved portion is steeper for larger values of b while the y-intercept does not change.

67. a)

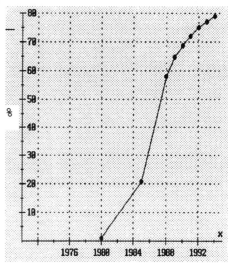

b) VCR sales started slowly then took off. Graph levels off as most households have one.

c) The limit would be 100%, but there will always be households without VCR's so this limit will never be reached.

d) Using calculator:

$y \approx \dfrac{77.8}{1 + 63.4e^{(-0.638x)}}$
where x is years after 1980.

69. a)

b) Data shows a fairly flat
   growth rate then levels off.

c) Natural limit would be 100%
   but will never be reached as
   some women will not be
   working at any given time.

d) $y \approx \dfrac{74.5}{1 + 3.14e^{(-0.08x)}}$,

   where x is years after 1960.

Section 6.6

1. a) $\log 2 + \log x = \log (2x)$

    b) $\log x^2 = 2 \log x$

    c) $\log\left(\dfrac{2}{x}\right) = \log 2 - \log x$

    d) $\log 2^x = x \log 2$

    e) $\log x - \log 2 = \log\left(\dfrac{x}{2}\right)$

3. Property #1
    $\log 6 = \log (2 \cdot 3) = \log 2 + \log 3$

5. Property #2
    $\log 9 = \log\left(3^2\right) = 2 \log 3$

7. $\log \sqrt{x} = \log\left(x^{1/2}\right) = \dfrac{1}{2} \log x$

9. $\log_b a \overset{?}{=} \dfrac{\ln a}{\ln b}$

    let $x = \log_b a$

    $b^x = a$

    $\ln b^x = \ln a$

    $x \ln b = \ln a$

    $x = \dfrac{\ln a}{\ln b}$

    $\log_b a = \dfrac{\ln a}{\ln b}$

11. $10^{2.6} = 3.98 \times 10^2$

    $10^2 \cdot 10^{0.6} = 3.98 \times 10^2$

    $10^{0.6} = 3.98$

    $\log 3.98 = 0.6$

13. $10^{-9.2} = 6.31 \cdot 10^{-10}$

    $10^{-10} \cdot 10^{0.8} = 6.31 \cdot 10^{-10}$

    $10^{0.8} = 6.31$

    $\log 6.31 = 0.8$

15. a) $\log (x + 1) + \log (x - 1)$

    $= \log[(x + 1)(x - 1)] = \log\left(x^2 - 1\right)$

    b) $\log\left(x^2 - 1\right) - \log(x - 1)$

    $= \log\left[\dfrac{(x^2 - 1)}{(x - 1)}\right]$

    $= \log\left[\dfrac{(x + 1)(x - 1)}{(x - 1)}\right] = \log (x + 1)$

17. a) $\log (x^2 + 3x + 2) - \log (x + 1)$

    $= \log\left[\dfrac{(x^2 + 3x + 2)}{(x + 1)}\right]$

    $= \log\left[\dfrac{(x + 1)(x + 2)}{(x + 1)}\right] = \log(x + 2)$

    b) $\log(x + 3) + \log(x - 2)$

    $= \log[(x + 3)(x - 2)] = \log\left(x^2 + x - 6\right)$

19. Same formula.

    $\log \dfrac{1}{\left[H^+\right]} = \log\left[H^+\right]^{-1}$

    $= -1 \cdot \log\left[H^+\right] = -\log\left[H^+\right]$

21. $\dfrac{1}{2} P = P e^{-rt}$

    $\dfrac{1}{2} = e^{-rt}$

    $t = \dfrac{\ln\left(\dfrac{1}{2}\right)}{-r} = -\dfrac{\ln\left(\dfrac{1}{2}\right)}{r} = \dfrac{-\ln\left(\dfrac{1}{2}\right)}{r}$

    $t = \dfrac{\ln\left(\dfrac{1}{2}\right)^{-1}}{r} = \dfrac{\ln 2}{r}$

23. a) $5^x = 20$

$\log 5^x = \log 20$

$x \log 5 = \log 20$

$x = \dfrac{\log 20}{\log 5} \approx 1.8614$

 b) $5^x = 20$

$\log_5 20 = x$

$x = \dfrac{\log 20}{\log 5} \approx 1.8614$

25. a) $3^{x+2} = 48$

$\log 3^{x+2} = \log 48$

$(x + 2) \log 3 = \log 48$

$(x + 2) = \dfrac{\log 48}{\log 3}$

$x = \dfrac{\log 48}{\log 3} - 2 \approx 1.5237$

 b) $3^{x+2} = 48$

$x + 2 = \log_3 48$

$x = \dfrac{\log 48}{\log 3} - 2 \approx 1.5237$

27. a) $4^{x-1} = 28$

$\log 4^{x-1} = \log 28$

$(x - 1) \log 4 = \log 28$

$x - 1 = \dfrac{\log 28}{\log 4}$

$x = \dfrac{\log 28}{\log 4} + 1 \approx 3.4037$

 b) $4^{x-1} = 28$

$x - 1 = \log_4 28$

$x = \dfrac{\log 28}{\log 4} + 1 \approx 3.4037$

29. $\log_2 x^2 + \log_2 x = 6$

$\log_2 x^3 = 6$

$2^6 = x^3$

$2^{6/3} = x$

$x = 2^2 = 4$

31. $\log x + \log x = 2$

$\log x^2 = 2$

$x^2 = 10^2$

$x = 10$

33. $S_{50} = \dfrac{1\left(1 - 1.10^{50}\right)}{(1 - 1.10)} \approx \$1163.91$

$S_{50} = \dfrac{50}{2}(1 + 50) = \$1275$

35. $S_{20} = \dfrac{108\left(1 - (2/3)^{20}\right)}{\left(1 - (2/3)\right)} \approx 323.9$ inches

37. $a_1 = 135, \quad r = \dfrac{2}{3}$

$S_{20} = \dfrac{135\left(1 - (2/3)^{20}\right)}{\left(1 - (2/3)\right)} \approx 404.88$ inches

39. $a_1 = 1, \quad r = \dfrac{1}{2}$

$S_\infty = \dfrac{1}{(1 - 1/2)} = 2$

41. $a_1 = 18, \quad r = \dfrac{5}{6}$

$S_\infty = \dfrac{18}{(1 - 5/6)} = \dfrac{18}{1/6} = 18 \cdot 6 = 108$ ft

Chapter Six Review

1. $r = 2$, $a_1 = 1/4$

$a_n = \dfrac{1}{4}(2)^{n-1}$

$f(x) = 0.125 \cdot 2^x$

$\dfrac{1}{4}(2)^{n-1} = \dfrac{1}{4}(2)^n \cdot 2^{-1} = \dfrac{1}{4} \cdot \dfrac{1}{2} \cdot 2^n$

$= \dfrac{1}{8}(2)^n = 0.125 \cdot 2^n$

3. $r = 2$, $a_1 = 1/16$

$a_n = \dfrac{1}{16}(2)^{n-1}$

$f(x) = 0.03125 \cdot 2^x$

$\dfrac{1}{16}(2)^{n-1} = \dfrac{1}{16}(2)^n \cdot 2^{-1} = \dfrac{1}{16} \cdot \dfrac{1}{2} \cdot 2^n$

$= \dfrac{1}{32}(2)^n = 0.03125 \cdot 2^n$

5. $r = 3$, $a_1 = 9$

$S_{10} = \dfrac{9\left(1 - 3^{10}\right)}{1 - 3} = 265{,}716$

$r > 1$ infinite sum is not appropriate.

7. $r = \dfrac{1}{2}$, $a_1 = 64$

$S_{10} = \dfrac{64\left(1 - \left(\dfrac{1}{2}\right)^{10}\right)}{1 - \dfrac{1}{2}} = 127.875$

$S_\infty = \dfrac{64}{1 - \dfrac{1}{2}} = \dfrac{64}{\dfrac{1}{2}} = 64 \cdot 2 = 128$

9. a) $2^{x-3} = 2^x \cdot 2^{-3} = \dfrac{2^x}{2^3} = \dfrac{2^x}{8}$

b) $3^{x+1} = 3^x \cdot 3^1 = 3 \cdot 3^x$

c) $3^{x-3} = 3^x \cdot 3^{-3} = \dfrac{3^x}{3^3} = \dfrac{3^x}{27}$

d) $2^{x+4} = 2^x \cdot 2^4 = 16 \cdot 2^x$

11.

$2^x = 16$	$\log_2 16 = x$	$x = \dfrac{\log 16}{\log 2} = 4$
$x^2 = 25$	$\log_x 25 = 2$	$x^2 = 5^2$ $x = 5$
$3^x = 81$	$\log_3 81 = x$	$x = \dfrac{\log 81}{\log 3} = 4$
$10^{1/2} = x$	$\log_{10} x = \dfrac{1}{2}$	$x = \sqrt{10} \approx 3.162$
$10^x = 19$	$\log_{10} 19 = x$	$x \approx 1.2788$
$4^0 = x$	$\log_4 x = 0$	$x = 1$

13.

$10^3 = x$	antilog $3 = x$	$\log x = 3$	$x = 1000$
$10^{1/2} = x$	antilog $\dfrac{1}{2} = x$	$\log x = \dfrac{1}{2}$	$x \approx 3.162$

15. $A = \$1{,}000\left(1 + \dfrac{0.06}{4}\right)^{4 \cdot 2} \approx \$1126.49$

$2{,}000 = 1{,}000\left(1 + \dfrac{0.06}{4}\right)^{4 \cdot t}$

$2 = \left(1 + \dfrac{0.06}{4}\right)^{4 \cdot t}$, $\log 2 = \log(1.015)^{4t}$

$\log 2 = 4t \log(1.015)$

$t = \dfrac{\log 2}{4 \log(1.015)} \approx 11.64$ yrs

17. $A = \$1{,}000\left(1 + \dfrac{0.06}{365}\right)^{365 \cdot 2} \approx \$1127.49$

$2{,}000 = 1{,}000\left(1 + \dfrac{0.06}{365}\right)^{365 \cdot t}$

$2 = \left(1 + \dfrac{0.06}{365}\right)^{365 \cdot t}$

$\log 2 = \log\left(1 + \dfrac{0.06}{365}\right)^{365 \cdot t}$

$\log 2 = 365t \log\left(1 + \dfrac{0.06}{365}\right)$

$t = \dfrac{\log 2}{365 \log\left(1 + \dfrac{0.06}{365}\right)} \approx 11.55$ yrs

19. $\dfrac{69}{5} = 13.8\%$

   $2 = e^{0.138t}, \quad \ln 2 = 0.138t$

   $t = \dfrac{\ln 2}{0.138} \approx 5$

21. $\log 3000 = \log(30 \cdot 100) = 3.47712$

   $\log 30 + \log 100 = 3.47712$

   $\log 30 + 2 = 3.47712$

   $\log 30 = 3.47712 - 2 = 1.47712$

23. $\log\left(\dfrac{x}{10}\right) = \log x - \log 10 = \log x - 1$

   so $\log x$ is 1 greater than $\log\left(\dfrac{x}{10}\right)$

25.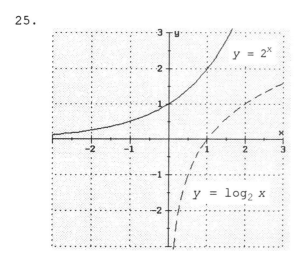

27. Intersect at $x = 0$, $(0,1)$, because for all $b \neq 0$, $b^0 = 1$.

29. Graph is a straight line.

31. $10^x = 0.1, \quad 10^x = 10^{-1}, \quad x = -1$

33. $36 = 10^x, \quad \log 36 = x$

   $x \approx 1.556$

35. $10^x = 0.75, \quad \log 0.75 = x$

   $x \approx -0.125$

37. $e^x = \dfrac{1}{e^2}, \quad e^x = e^{-2}, \quad x = -2$

39. $\left(\dfrac{1}{e}\right)^x = e, \quad \left(e^{-1}\right)^x = e$

   $e^{-x} = e^1, \quad -x = 1, \quad x = -1$

41. $4^{x+1} = 32, \quad \left(2^2\right)^{x+1} = 2^5$

   $2^{2x+2} = 2^5, \quad 2x + 2 = 5$

   $2x = 3, \quad x = \dfrac{3}{2}$

43. $3^{2x} = 6, \quad \log_3 6 = 2x$

   $x = \dfrac{1}{2}\left(\dfrac{\log 6}{\log 3}\right) \approx 0.815$

45. $\log x = -1, \quad 10^{-1} = x$

   $x = 0.1$

47. $\log_2 x = 1, \quad 2^1 = x, \quad x = 2$

49. $\log_2 2 = x, \quad x = 1$

51. $\log 10{,}000 = x, \quad 10^x = 10{,}000$

   $10^x = 10^4, \quad x = 4$

53. $\log x = 2, \quad 10^2 = x, \quad x = 100$

55. $\log_3 9 = x, \quad x = \dfrac{\log 9}{\log 3} = 2$

   or $3^x = 9, \quad 3^x = 3^2, \quad x = 2$

57. $\log_{27} 9 = x \quad x = \dfrac{\log 9}{\log 27} = \dfrac{2}{3}$

   or $27^x = 9, \quad 3^{3x} = 3^2$

   $3x = 2, \quad x = \dfrac{2}{3}$

59. $\ln x = 3 \quad x = e^3 \approx 20.086$

61. $\ln x = \dfrac{1}{e}$ $x = e^{\left(\frac{1}{e}\right)} \approx 1.445$

63. $a_1 = \$0.01$ $r = 2$

   a) $a_{10} = \$0.01 \cdot 2^{10-1} = \$5.12$

   b) $a_{20} = \$0.01 \cdot 2^{20-1} = \$5242.88$

   c) $S_{20} = \dfrac{\$0.01\left(1 - 2^{20}\right)}{1 - 2} = \$10{,}485.75$

   d)

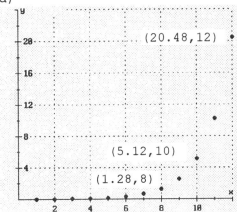

   e) Yes!

   f) $S_{52} = \dfrac{\$0.01\left(1 - 2^{52}\right)}{1 - 2} \approx \$4.5 \cdot 10^{13}$

65. $\dfrac{40}{12} \approx 3\dfrac{1}{3}$ doubles

   $\$1, \quad \$2, \quad \$4, \quad \$8, \approx \$10$

   1960 1972 1984 1996 2000

   $A = Pe^{kt}$

   $2 = e^{k \cdot 12}, \quad \ln 2 = 12k$

   $k = \dfrac{\ln 2}{12} \approx 0.05776$

   $A = \$1.00e^{0.05776t}$

67. $P \approx 57403(0.9921)^x$

   $r^* = 0.9921 - 1 = -0.0079 \Rightarrow -0.79\%$

   $P_{2000} \approx 57403(0.9921)^{70} \approx 33{,}000$

   Estimate seems reasonable,
   matches population decline.

69. a)

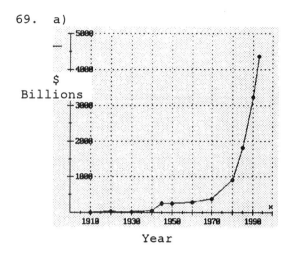

Year

b) $y \approx \left(4.1986 \times 10^9\right)(1.0868)^x$

c) Wars; spending in the 1980's.

d) Answer will vary with the
year.

e)

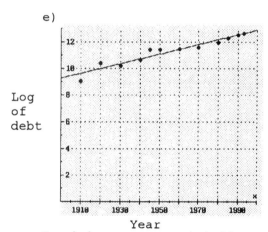

Year

Graph becomes somewhat linear.

71. a) Deaths $\approx 8715(1.254)^x$

new cases $\approx 11491(1.263)^x$

b) Deaths $\approx 8715(1.254)^{15} \approx 260,000$
new cases
$\approx 11491(1.263)^{15} \approx 381,000$

c) ≈ 2001

d) Yes, estimates are low. The
number of new cases is
growing faster than earlier
data suggests.

Chapter Six Test

1. $3 \diagdown_3 \diagup 6 \diagdown_3 \diagup 9 \diagdown_3 \diagup 12 \diagdown_3 \diagup 15$

   a) Arithmetic

   b) $a_n = 3 + (n - 1)3 = 3 + 3n - 3$

       $a_n = 3n$

       $f(x) = 3x$

2. $\dfrac{1}{8}, \dfrac{1}{4}, \dfrac{1}{2}, 1, 2 ; \quad r = 2$

   a) Geometric

   b) $a_n = \left(\dfrac{1}{8}\right) \cdot 2^{n-1}$

       $f(x) = 0.0625 \cdot 2^x$

       $\left(\dfrac{1}{8}\right) \cdot 2^{n-1} = \left(\dfrac{1}{8}\right) \cdot 2^n \cdot 2^1$

       $= \left(\dfrac{1}{8}\right)\left(\dfrac{1}{2}\right)2^n = \dfrac{1}{16} \cdot 2^n$

       $= 0.0625 \cdot 2^n$

   d) $S_{10} = \dfrac{\frac{1}{8}\left(1 - 2^{10}\right)}{1 - 2} = 127.875$

3. $\dfrac{1}{4}, \dfrac{1}{8}, \dfrac{1}{16}, \dfrac{1}{32} ; \quad r = \dfrac{1}{2}$

   a) Geometric

   b) $a_n = \left(\dfrac{1}{4}\right)\left(\dfrac{1}{2}\right)^{n-1}$

       $f(x) = 0.5 \cdot 0.5^x$

       $\left(\dfrac{1}{4}\right)\left(\dfrac{1}{2}\right)^{n-1} = \left(\dfrac{1}{4}\right)\left(\dfrac{1}{2}\right)^n\left(\dfrac{1}{2}\right)^{-1}$

       $= \left(\dfrac{1}{4}\right)(2)\left(\dfrac{1}{2}\right)^n = \dfrac{1}{2} \cdot \left(\dfrac{1}{2}\right)^n$

       $= 0.5 \cdot 0.5^n$

3. d) $S_{10} = \dfrac{\frac{1}{4}\left(1 - \left(\frac{1}{2}\right)^{10}\right)}{1 - \frac{1}{2}} \approx 0.4995$

   e) $S_\infty = \dfrac{\left(\frac{1}{4}\right)}{1 - \frac{1}{2}} = \dfrac{1}{4} \cdot 2 = 0.5$

4. $3 \diagdown_9 \diagup 12 \diagdown_{15} \diagup 27 \diagdown_{21} \diagup 48 \diagdown_{27} \diagup 75$
   $\qquad\quad {}_6 \qquad {}_6 \qquad {}_6$

   a) Quadratic

   c) $f(x) = 3x^2$

5. $y = 40.4x + 16093$

   $f(x) = 16093 \cdot 1.0025^x$

   $40.4(60) + 16093 = 18517$

   $16093 \cdot 1.0025^{60} \approx 18694$

6. a)

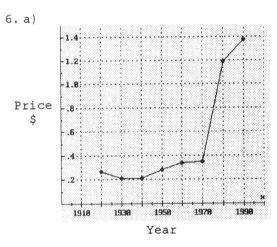

Year

   b) Answers will vary, could be exponential or logistic.

   c) Using exponential model:
      $f(x) \approx 0.161(1.0263)^x$

7. a) $10^x = 3$, $\log 3 = x \approx 0.477$

b) $5^{2x} = 0.25$, $2x = \log_5 0.25$

$$x = \left(\frac{1}{2}\right)\left(\frac{\log 0.25}{\log 5}\right) \approx -0.431$$

c) $10^x = -1$ $\{\}$

d) $e^x = \frac{1}{e}$, $e^x = e^{-1}$, $x = -1$

e) $5^{x-2} = \frac{1}{125}$, $5^{x-2} = 5^{-3}$

$x - 2 = -3$, $x = -1$

f) $\log_2 x = -2$, $2^{-2} = x$, $x = \frac{1}{4}$

g) $\log_5 x = 3$, $5^3 = x$, $x = 125$

h) $\log_x 32 = 5$, $x^5 = 32$, $x^5 = 2^5$,

$x = 2$

i) $\log x = 1$, $10^1 = x$, $x = 10$

j) $\ln x = 1.5$, $e^{1.5} = x$, $x \approx 4.482$

8. a) $\log_n 1 = 0$

b) $\log 1 = 0$

c) $\log_2 x + \log_2(x + 1) = \log_2\big(x(x + 1)\big)$

$= \log_2\big(x^2 + x\big)$

d) $\log_3\big(x^2 - 9\big) - \log_3(x + 3) = \log_3\left(\frac{x^2 - 9}{x + 3}\right)$

$= \log_3\left(\frac{(x + 3)(x - 3)}{(x + 3)}\right) = \log_3(x - 3)$

9. $y = b^{0+2}$, $y = b^2$

$y = b^{x+2}$, $y = b^x \cdot b^2$, $y = b^x b^2$

10.

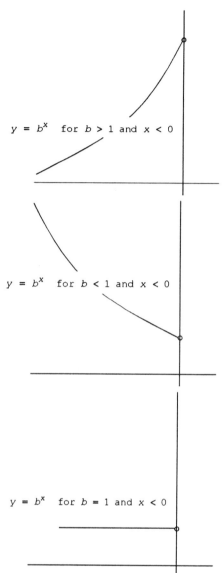

$y = b^x$ for $b > 1$ and $x < 0$

$y = b^x$ for $b < 1$ and $x < 0$

$y = b^x$ for $b = 1$ and $x < 0$

Note: the point on the y-axis is <u>open</u>.

11. Evaluate $\left(1 + \frac{0.06}{12}\right)^{12}$ and

compare to $1 + r^{*}$

$1 + r^{*} = \left(1 + \frac{0.06}{12}\right)^{12} \approx 1.0617$

$r^{*} \approx 0.0617$

12. Changes a in $y = ab^x$

13. $\log(100x) = \log 100 + \log x$

    $\log 100 = 2$

    so $\log(100x) = 2 + \log x$.

14. Answers will vary, examples are
    (0,1) & (1,0)
    (1,10) & (10,1)
    By definition
    $y = 10^x$ is $x = \log y$;
    $y = \log x$ has reversed coordinates.

15. a) $pH = -\log(5.01 \times 10^{-9}) \approx 8.3$

    b) $pH = -\log[H^+] = 11.9$

    $-11.9 = \log[H^+],\quad 10^{-11.9} = [H^+]$

    $[H^+] \approx 1.26 \cdot 10^{-12} M$

16.

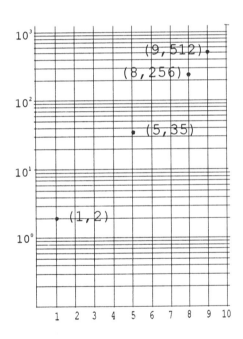

17. $y = 1$ when $x = 0$, $y = b^x$

    $y = 3$ when $x = 1$, $y = 3 = b^1$

    $y = 3^x$

18. $x = 1$ when $y = 0$, $x = b^y$, $1 = b^0$

    $x = 2$ when $y = 1$, $2 = b^1$, $b = 2$

    $f(x) = \log_2 x$

Section 7.0

1. $w = 2c, \quad s = 3c, \quad w + s + c = 60$

    $2c + 3c + c = 60$

    $6c = 60$

    $c = 10\,\text{hrs}$

    $w = 2(10) = 20\,\text{hrs}$

    $s = 3(10) = 30\,\text{hrs}$

    $20 = 2(10)\,\checkmark$

    $30 = 3(10)\,\checkmark$

    $20 + 30 + 10 = 60\,\checkmark$

3.

    $c = 2p, \quad c + f + p = 28$

    $4c + 9f + 4p = 222$

    $2p + f + p = 28, \quad 3p + f = 28$

    $4(2p) + 9f + 4p = 222$

    $\big(-4(3p + f) = 28(-4)\big) = -12p - 4f = -112$

    $ 12p + 9f = 222$

    $ 5f = 110$

    $ f = 22$

    $3p + 22 = 28$

    $3p = 6$

    $p = 2$

    $c = 2(2) = 4$

    $4 = 2(2)\,\checkmark$

    $4 + 22 + 2 = 28\,\checkmark$

    $4(4) + 9(22) + 4(2) = 222$

    $16 + 198 + 8 = 222\,\checkmark$

5. $f = p, \quad c + f + p = 29$

    $4c + 9f + 4p = 121$

    $c + f + f = 29, \quad 4c + 9f + 4f = 121$

    $c + 2f = 29, \quad\quad 4c + 13f = 121$

    $\big(-4(c + 2f) = 29(-4)\big) = -4c - 8f = -116$

    $ = +4c + 13f = 121$

    $ 5f = 5$

    $ f = 1$

    $p = f \Rightarrow p = 1$

    $c + 1 + 1 = 29$

    $c = 27$

    $1 = 1\,\checkmark$

    $27 + 1 + 1 = 29\,\checkmark$

    $4(27) + 9(1) + 4(1) = 121$

    $108 + 9 + 4 = 121\,\checkmark$

7. $3x + y = 4$

    $-3x - 3x$

    $y = -3x + 4$

9. $x - 3y = 7$

    $+3y + 3y$

    $x = 3y + 7$

11. $4y - 2x = 5$

    $-4y - 4y$

    $-2x = -4y + 5$

    $\dfrac{-2x}{-2} = \dfrac{-4y}{-2} + \dfrac{5}{-2}$

    $x = 2y - \dfrac{5}{2}$

13. $4x - 2y = -3$

    $-4x - 4x$

    $-2y = -4x - 3$

    $\dfrac{-2y}{-2} = \dfrac{-4x}{-2} - \dfrac{3}{-2}$

    $y = 2x + \dfrac{3}{2}$

Section 7.0

15. $y = x^2 + 5$
    $\underline{-5 \qquad -5}$
    $y - 5 = x^2$ or
    $x^2 = y - 5$

17. $x - 3 = -2y^2 + x^2$
    $\underline{+2y^2 \qquad + 2y^2}$
    $2y^2 + x - 3 = x^2$
    $\underline{\quad - x + 3 \qquad - x + 3}$
    $2y^2 = x^2 - x + 3$
    $\dfrac{2y^2}{2} = \dfrac{x^2}{2} - \dfrac{x}{2} + \dfrac{3}{2}$
    $y^2 = \dfrac{x^2}{2} - \dfrac{x}{2} + \dfrac{3}{2}$

19. $x^2 + y^2 = 8$
    $\underline{-x^2 \qquad - x^2}$
    $y^2 = 8 - x^2$
    $\left(y^2\right)^{1/2} = \left(8 - x^2\right)^{1/2}$ or $\sqrt{y^2} = \sqrt{8 - x^2}$
    $y = \pm\sqrt{8 - x^2}$

21. $x^2 - y^2 = 10$
    $\underline{+y^2 \qquad\quad + y^2}$
    $x^2 = y^2 + 10$
    $\left(x^2\right)^{1/2} = \left(y^2 + 10\right)^{1/2}$
    or $\sqrt{x^2} = \sqrt{y^2 + 10}$
    $x = \pm\sqrt{y^2 + 10}$

23. $y = 3x + 4$
    $3y - 2x = 9$

    $3(3x - 4) - 2x = 9$
    $9x - 12 - 2x = 9$
    $7x - 12 = 9$
    $\underline{+12 \qquad\quad + 12}$
    $7x = 21$
    $x = 3$
    $y = 3(3) - 4$
    $y = 9 - 4 = 5$
    $(3, 5)$
    $5 = 3(3) - 4$
    $5 = 9 - 4$ ✓
    $3(5) - 2(3) = 9$
    $15 - 6 = 9$ ✓

25. $0.6x + y = 3$
    $1.4x + y + 3 = 0$
    $\underline{\qquad\qquad - 3 \quad - 3}$
    $1.4x + y = -3 \Rightarrow \qquad 1.4x + y = -3$
    $-(0.6x + y = 3) \Rightarrow \underline{-0.6x - y = -3}$
    $\qquad\qquad\qquad\qquad 0.8x \qquad\quad = -6$
    $x = -7.5$
    $1.4(-7.5) + y = -3$
    $-10.5 + y = -3$
    $y = 7.5$
    $(-7.5, 7.5)$
    $0.6(-7.5) + 7.5 = 3$
    $-4.5 + 7.5 = 3$ ✓
    $1.4(-7.5) + 7.5 + 3 = 0$
    $-10.5 + 10.5 = 0$ ✓

27. $x + 2y = 9.4 \Rightarrow -3(x + 2y) = 9.4(-3)$

$3x - 5y = 4 \qquad -3x - 6y = -28.2$

$\underline{-3x - 6y = -28.2}$

$\qquad -11y = -24.2$

$y = 2.2$

$x + 2(2.2) = 9.4$

$x + 4.4 = 9.4$

$x = 5$

$(5, 2.2)$

$5 + 2(2.2) = 9.4$

$5 + 4.4 = 9.4 \checkmark$

$3(5) - 5(2.2) = 4$

$15 - 11 = 4 \checkmark$

29. $y = 1.8x - 6.6$

$y + 1.4x = 3$

$(1.8x - 6.6) + 1.4x = 3$

$3.2x - 6.6 = 3$

$3.2x = 9.6$

$x = 3$

$y = 1.8(3) - 6.6$

$y = 5.4 - 6.6$

$y = -1.2$

$(3, -1.2)$

$-1.2 = 1.8(3) - 6.6$

$-1.2 = 5.4 - 6.6 \checkmark$

$-1.2 + 1.4(3) = 3$

$-1.2 + 4.2 = 3 \checkmark$

31. $5x = 14.7 + 6y$

$5x - 6y = 14.7$

$\underline{2x + 6y = -4.2}$

$7x = 10.5$

$x = 1.5$

$5(1.5) = 14.7 + 6y$

$7.5 - 14.7 = 6y$

$6y = -7.2$

$y = -1.2$

$(1.5, -1.2)$

$5(1.5) = 14.7 + 6(-1.2)$

$7.5 = 14.7 - 7.2 \checkmark$

$2(1.5) + 6(-1.2) = -4.2$

$3 - 7.2 = -4.2 \checkmark$

33. $acx + bcy = ce$

$\underline{-(acx + ady = af)}$

$(bc - ad)y = ce - af$

$y = \dfrac{ce - af}{bc - ad}$

or

$acx + ady = af$

$\underline{-(acx + bcy = ce)}$

$(ad - bc)y = af - ce$

$y = \dfrac{af - ce}{ad - bc}$

35. If one variable has a coefficient of 1 and is easy to solve for, then substitution may be most practical.

37. $2a + 3b + 2c = 1$

 $3a - b + 4c = 13$

 $a + b = 1$

 $a = 1 - b$

 $2(1 - b) + 3b + 2c = 1$

 $2 - 2b + 3b + 2c = 1$

 $b + 2c = -1$

 $b = -1 - 2c$

 $3(1 - b) - b + 4c = 13$

 $3 - 3b - b + 4c = 13$

 $-4b + 4c = 10$

 $-4(-1 - 2c) + 4c = 10$

 $4 + 8c + 4c = 10$

 $12c = 6$

 $c = \dfrac{1}{2}$

 $b = -1 - 2\left(\dfrac{1}{2}\right) = -1 - 1 = -2$

 $a = 1 - (-2) = 1 + 2 = 3$

 $2(3) + 3(-2) + 2\left(\dfrac{1}{2}\right) = 1$

 $6 - 6 + 1 = 1 \checkmark$

 $3(3) - (-2) + 4\left(\dfrac{1}{2}\right) = 13$

 $9 + 2 + 2 = 13 \checkmark$

 $3 + (-2) = 1 \checkmark$

39. $2a + b - 4c = 17$

 $5a - 2b + c = 25$

 $4b + 3C + 10 = 0 \ , \ 4b + 3c = -10$

 $4b = -3c - 10$

 $b = \dfrac{-3}{4}c - \dfrac{10}{4}$

 $2a + \left(-\dfrac{3}{4}c - \dfrac{5}{2}\right) - 4c = 17$

 $5a - 2\left(-\dfrac{3}{4}c - \dfrac{5}{2}\right) + c = 25$

 $2a - \dfrac{3}{4}c - 4c = 17 + \dfrac{5}{2}$

 $2a - 4.75c = 19.5$

 $5a + 1.5c + 5 + c = 25$

 $5a + 2.5c = 20$

 $5(2a - 4.75c) = 19.5(5)$

 $10a - 23.75c = 97.5$

 $-2(5a + 2.5c) = 20(-2)$

 $-10a - 5c = -40$

 $\underline{10a - 23.75c = 97.5}$

 $ - 28.75c = 57.5$

 $c = -2$

 $4b + 3(-2) + 10 = 0$

 $4b - 6 + 10 = 0$

 $4b + 4 = 0$

 $4b = -4$

 $b = -1$

 $2a + (-1) - 4(-2) = 17$

 $2a - 1 + 8 = 17$

 $2a + 7 = 17$

 $2a = 10$

 $a = 5$

 $2(5) + (-1) - 4(-2) = 17$

 $10 - 1 + 8 = 17$

 $5(5) - 2(-1) + (-2) = 25$

 $25 + 2 - 2 = 25 \checkmark$

 $4(-1) + 3(-2) + 10 = 0$

 $-4 - 6 + 10 = 0$

Section 7.0

41. Grams of carbohydrate,
    Grams of fat, and
    Grams of protein.

43. No, vertical lines (x=c)
    cannot.

45. $x = 2 \Rightarrow 1x + 0y = 2$

    $a = 1$, $b = 0$ and $c = 2$

47. Slope at
    $k = -0.04 = 2a(0) + b$

    $b = -0.04 = -\dfrac{4}{100} = -\dfrac{1}{25}$

    At k, $700 = a(0)^2 + (-0.04)(0) + c$

    $c = 700$

    At L, $x = 2000$, slope, $m = 0.03$

    $0.03 = 2a(2000) + (-0.04)$

    $0.07 = 4000a$

    $a = \dfrac{0.07}{4000} = \dfrac{7}{400000}$

    Equation for the curve is:

    $$y = \frac{7}{400000} x^2 - \frac{1}{25} x + 700$$

49. At M, slope, $m = 2a(0) + b = -0.05$

    $b = -0.05 = -\dfrac{5}{100} = -\dfrac{1}{20}$

    At N, slope, $m = 0.06$
    $= 2a(2400) - 0.05$

    $0.11 = 4800a$

    $a = \dfrac{0.11}{4800} = \dfrac{11}{480000}$

    At N,

    $1200 = \dfrac{11}{480,000} (2400)^2 - \dfrac{1}{20} (2400) + c$

    $1200 = 132 - 120 + c$

    $1188 = c$

    Equation for the curve is:

    $$y = \frac{11}{480000} x^2 - \frac{1}{20} x + 1188$$

1.

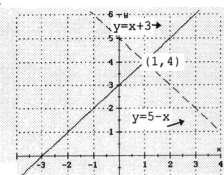

3. $2x - 3 = 2y$

$$y = x - \frac{3}{2}$$

$x - y = 1.5$

$y = x - 1.5$

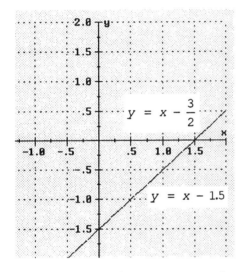

Coincident lines; same slope and y-intercept.

$$x - \frac{3}{2} = x - 1.5$$

$$-\frac{3}{2} = -1.5 \text{ True result}$$

5. $2y = 1.5x + 1$

$y = 0.75x + 0.5$

$4y - 3x + 8 = 0$

$4y = 3x - 8$

$y = 0.75x - 2$

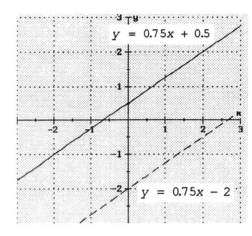

Note: Same slope, but different y-intercepts creates parallel lines.

$0.75x + 0.5 = 0.75x - 2$

$0.5 = -2$ False result

7. $x + 2y = 4$

$2y = -x + 4$

$$y = -\frac{x}{2} + 2$$

$x - y = 3$

$y = x - 3$

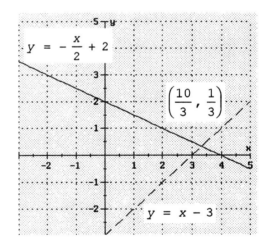

Section 7.1

9. $2x - y = 4$
    $y = 2x - 4$
    $4x + y = -1$
    $y = -4x - 1$

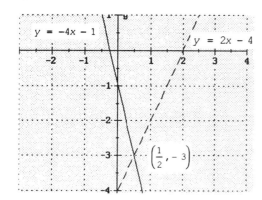

11. $y = \dfrac{x}{3} + 2$

    $3y - 6 = x$

    $3y = x + 6$

    $y = \dfrac{x}{3} + 2$

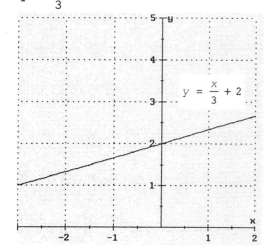

Coincident lines: again same slope & y-intercept.

$\dfrac{x}{3} + 2 = \dfrac{x}{3} + 2$

$2 = 2$ true result

13. $a = 0$: $0x + by = c$, $y = \dfrac{c}{b}$

    horizontal line (*b*)

15. $c = 0$; $ax + by = 0$, $by = -ax$, $y = \dfrac{-a}{b} x$

    $slope = -\dfrac{a}{b}$,   y – intercept $(0, 0)$

    (*c*)

17. $a = 1$; $1x + by = c$

    $by = c - x$, $y = -\dfrac{1}{b} x + \dfrac{c}{b}$

    slope $= -\dfrac{1}{b}$,   y – intercept $\left(0, \dfrac{c}{b}\right)$

    (*d*)

19. Quantity: $1600 & $12,000
    investments.
    Value: 5% and 8% annual
    interest.

21. Quantity: 20 hrs & 15 hrs
    worked.
    Value: $6.50 & $7.25 per hour.

23. Quantity: 100 ml & 1000 ml.
    Value: 8% and 0%.

Answers for #25 and #27 may vary-
possible answers are:

25. Question: What is the total
    value of the money?
    Facts: Dimes are worth $0.10
    each. Quarters are worth $0.25
    each.

27. Question: What is the measure
    of each angle?
    Facts: The sum of the interior
    angles in a triangle is 180°.

29.

	Qty (Cr.-hrs)	Value (pts/cr)	Qty · Value (points)
Computer Science	4	(A) 4	16
Psycology	3	(B) 3	9
English	3	(C) 2	6
P.E.	1	(B) 3	3
Total	11		34

GPA $= \dfrac{34}{11} \approx 3.09$ points/credit

Section 7.1

31.

	Qty (Credits)	Value (pts/cr)	Qty · Value (points)
	75	3.08	231
	x	4	4x
Total	75+x	3.25 (Desired)	4x+231

$231 + 4x = 3.25(75 + x)$

$231 + 4x = 243.75 + 3.25x$

$0.75x = 12.75$

$x = 17$
Student needs 17 credits of
"A"'s

33.

Dog Food	Qty (lbs)	Value (% Protein)	Qty · Value (lbs protein)
A	x	0.15	0.15x
B	y	0.10	0.10y
Total	1000	0.12	1000(0.12)=120

$x + y = 1,000, \quad x = 1000 - y$

$0.15x + 0.10y = 120$

$0.15(1,000 - y) + 0.10y = 120$

$150 - 0.15y + 0.10y = 120$

$-0.05y = -30$

$y = 600$

$x + 600 = 1000$

$x = 400$

Needs 400 lbs of A & 600 lbs of
B.

35.

	Qty ($)	Value (% interest)	Qty · Value ($ interest)
	x	0.06	0.06x
	y	0.09	0.09y
Total	$75,000	0.08	$6,000

$x + y = 75,000 \quad x = 75,000 - y$

$0.06x + 0.09y = 6,000$

$(0.06)(75000 - y) + 0.09y = 6000$

$4500 - 0.06y + 0.09y = 6000$

$0.03y = 1500$

$y = 50,000$

$x = 75,000 - 50,000 = 25,000$

Invest $25,000 at 6% and
$50,000 at 9%.

37.

	Qty (liters)	Value (% glucose)	Qty · Value (ltrs glucose)
	x	0	0
	y	0.5	0.5y
Total	0.5	0.05	0.025

$x + y = 0.5$

$0.5y = 0.025; \quad y = 0.05$

$x + 0.05 = 0.5$

$x = 0.45$

need 0.454 L distilled water
0.05 L 50% glucose.

39. Pkg 1: $y = 0.10x + 250$

Pkg 2: $y = 0.08x + 350$

$0.10x + 250 = 0.08x + 350$

$0.02x = 100$

$x = 5000$
Packages are equal when sales
are $5000.

41. Pkg 3: $y = 0.04x + 550$

Pkg 4: $y = 0.04x + 750$
Note - parallel lines
Package 4 is always better.

43. Pkg 1: $y = 0.10x + 250$

Pkg 5: $y = 0.06x + 550$

$0.10x + 250 = 0.06x + 550$

$0.04x = 300$

$x = 7500$
Packages are equal when sales
are $7500.

45. Increasing % of sales changes
the slope of the graph. The
graph becomes steeper.

47. Write equations for parallel
lines, use same slope but
different y-intercepts.

Section 7.2

1. $\begin{bmatrix} 1 & 2 \\ 4 & 4 \end{bmatrix} \cdot \begin{bmatrix} x \\ y \end{bmatrix} = \begin{bmatrix} 5 \\ 6 \end{bmatrix}$

   $\begin{bmatrix} x \\ y \end{bmatrix} = \begin{bmatrix} -2 \\ 3.5 \end{bmatrix}$

   $x = -2, \quad y = 3.5$

3. $\begin{bmatrix} 1 & 2 \\ 2 & 4 \end{bmatrix} \cdot \begin{bmatrix} x \\ y \end{bmatrix} = \begin{bmatrix} 5 \\ 6 \end{bmatrix}$

   $\det[A] = 0$

   $x + 2y = 5$

   $2y = -x + 5$

   $y = -\dfrac{1}{2}x + 5$

   $2x + 4y = 6$

   $4y = -2x + 6$

   $y = -\dfrac{1}{2}x + \dfrac{3}{2}$

   Lines are parallel.

5. $\begin{bmatrix} 1 & 2 \\ 0 & 4 \end{bmatrix} \cdot \begin{bmatrix} x \\ y \end{bmatrix} = \begin{bmatrix} 5 \\ 6 \end{bmatrix}$

   $\begin{bmatrix} x \\ y \end{bmatrix} = \begin{bmatrix} 2 \\ 1.5 \end{bmatrix}$

   $x = 2, \quad y = 1.5$

7. $\begin{bmatrix} 1 & 2 \\ -2 & 4 \end{bmatrix} \cdot \begin{bmatrix} x \\ y \end{bmatrix} = \begin{bmatrix} 5 \\ 6 \end{bmatrix}$

   $\begin{bmatrix} x \\ y \end{bmatrix} = \begin{bmatrix} 1 \\ 2 \end{bmatrix}$

   $x = 1, \quad y = 2$

9. All contain the equation
   $1x + 2y = 5$ and the equation
   $ax + 4y = 6$. Changing a will
   change the slope of the equation.
   The y-intercept remains the same
   $(0, 1.5)$ for each value of a.

11. $5x = 28.2 + 2y$

    $5x - 2y = 28.2$

    $3x + 5y = -5.4$

    $\begin{bmatrix} 5 & -2 \\ 3 & 5 \end{bmatrix} \cdot \begin{bmatrix} x \\ y \end{bmatrix} = \begin{bmatrix} 28.2 \\ -5.4 \end{bmatrix}$

    $\begin{bmatrix} x \\ y \end{bmatrix} = \begin{bmatrix} 4.2 \\ -3.6 \end{bmatrix}$

    $x = 4.2, \quad y = -3.6$

13. $3y = 2x + 12, \quad y + 2 = \dfrac{2x}{3}$

    $3y - 2x = 12, \quad y - \dfrac{2x}{3} = -2$

    $\begin{bmatrix} 3 & -2 \\ 1 & -2/3 \end{bmatrix} \cdot \begin{bmatrix} y \\ x \end{bmatrix} = \begin{bmatrix} 12 \\ -2 \end{bmatrix}$

    $\det[A] = 0$

    $3y = 2x + 12, \quad y + 2 = \dfrac{2}{3}x$

    $y = \dfrac{2}{3}x + 4, \quad y = \dfrac{2}{3}x - 2$

    Parallel lines.

15. $x - 1.5y = 3.5, \quad 2x - 7 = 3y$

    $\qquad\qquad\qquad\quad 2x - 3y = 7$

    $\begin{bmatrix} 1 & -1.5 \\ 2 & -3 \end{bmatrix} \cdot \begin{bmatrix} x \\ y \end{bmatrix} = \begin{bmatrix} 3.5 \\ 7 \end{bmatrix}$

    $\det[A] = 0$

    $x - 1.5y = 3.5, \quad 2x - 7 = 3y$

    $x - 3.5 = 1.5y = \dfrac{3}{2}y, \quad y = \dfrac{2}{3}x - \dfrac{7}{3}$

    $y = \dfrac{2}{3}x - \dfrac{7}{3}$

    Coincident lines.

17.

	Qty	Value	Qty · Value
Honduran	x	18.70	18.70x
Indonesian	y	23.65	23.65y
Total	200	19.80	3960

$x + y = 200$

$18.70x + 23.65y = 3960$

$$\begin{bmatrix} 1 & 1 \\ 18.70 & 23.65 \end{bmatrix} \cdot \begin{bmatrix} x \\ y \end{bmatrix} = \begin{bmatrix} 200 \\ 3960 \end{bmatrix}$$

$$\begin{bmatrix} x \\ y \end{bmatrix} \approx \begin{bmatrix} 155.6 \\ 44.4 \end{bmatrix}$$

Need ≈ 156kg of Honduran
and ≈ 44kg of Indonesian.

19.

Qty	Value	Qty · Value
x	0.03	0.03x
y	0.20	0.20y
1000	0.064	64

$x + y = 1000$

$0.03x + 0.20y = 64$

$$\begin{bmatrix} 1 & 1 \\ 0.03 & 0.20 \end{bmatrix} \cdot \begin{bmatrix} x \\ y \end{bmatrix} = \begin{bmatrix} 1000 \\ 64 \end{bmatrix}$$

$$\begin{bmatrix} x \\ y \end{bmatrix} = \begin{bmatrix} 800 \\ 200 \end{bmatrix}$$

800 l of 3%, 200 l of 20%

21.

	Qty	Value	Qty · Value
water	x	0	0x
	y	0.20	0.20y
	1000	0.03	30

$x + y = 1000$

$0.20y = 30$

$y = 150$

$x + 150 = 1000$

$x = 850$

850 gal of water and 150 gal
of 20% hydrogen peroxide.

23. $9d + 13q = 3.5$

$12d + 9q = 3$

$$\begin{bmatrix} 9 & 13 \\ 12 & 9 \end{bmatrix} \cdot \begin{bmatrix} d \\ q \end{bmatrix} = \begin{bmatrix} 3.5 \\ 3 \end{bmatrix}$$

$$\begin{bmatrix} d \\ q \end{bmatrix} \approx \begin{bmatrix} 0.1 \\ 0.2 \end{bmatrix}$$

dimes ≈ 0.1oz
quarters ≈ 0.2oz

25. $0.045x + 0.085y = 1782$

$x + y = 23600$

$$\begin{bmatrix} 0.045 & 0.085 \\ 1 & 1 \end{bmatrix} \cdot \begin{bmatrix} x \\ y \end{bmatrix} = \begin{bmatrix} 1782 \\ 23600 \end{bmatrix}$$

$$\begin{bmatrix} x \\ y \end{bmatrix} \approx \begin{bmatrix} 5600 \\ 18000 \end{bmatrix}$$

$5600 at 4.5%
$18,000 at 8.5%

27. $[A]^{-1} = \begin{bmatrix} -2 & -3 \\ 1 & 1 \end{bmatrix}$

29. $[C]^{-1} = \begin{bmatrix} 0.2 & 0.2 \\ -0.1 & 0.4 \end{bmatrix}$

31. $\begin{bmatrix} 1 & 3 \\ -1 & -2 \end{bmatrix} \cdot \begin{bmatrix} x \\ y \end{bmatrix} = \begin{bmatrix} 1x + 3y \\ -1x - 2y \end{bmatrix}$

33. $\begin{bmatrix} 1 & 3 \\ -1 & -2 \end{bmatrix} \cdot \begin{bmatrix} 2 \\ -1 \end{bmatrix} = \begin{bmatrix} 2 - 3 \\ -2 + 2 \end{bmatrix} = \begin{bmatrix} -1 \\ 0 \end{bmatrix}$

35. $\begin{bmatrix} 4 & -2 \\ 1 & 2 \end{bmatrix} \cdot \begin{bmatrix} 2 \\ -1 \end{bmatrix} = \begin{bmatrix} 8 + 2 \\ 2 - 2 \end{bmatrix} = \begin{bmatrix} 10 \\ 0 \end{bmatrix}$

37. $\begin{bmatrix} 1 & 3 \\ -1 & -2 \end{bmatrix} \cdot \begin{bmatrix} 2 & 3 \\ 4 & 5 \end{bmatrix} = \begin{bmatrix} 2+12 & 3+15 \\ -2-8 & -3-10 \end{bmatrix} = \begin{bmatrix} 14 & 18 \\ -10 & -13 \end{bmatrix}$

39. $\begin{bmatrix} 2 & 3 \\ 4 & 5 \end{bmatrix}\begin{bmatrix} 1 & 3 \\ -1 & -2 \end{bmatrix} = \begin{bmatrix} 2-3 & 6-6 \\ 4-5 & 12-10 \end{bmatrix}$

$$= \begin{bmatrix} -1 & 0 \\ -1 & 2 \end{bmatrix}$$

41. $\begin{bmatrix} 4 & -2 \\ 1 & 2 \end{bmatrix}\begin{bmatrix} -2 & -1 \\ 1 & 2 \end{bmatrix} = \begin{bmatrix} -8-2 & -4-4 \\ -2+2 & -1+4 \end{bmatrix}$

$$= \begin{bmatrix} -10 & -8 \\ 0 & 3 \end{bmatrix}$$

43. $\begin{bmatrix} -2 & -1 \\ 1 & 2 \end{bmatrix}\begin{bmatrix} 4 & -2 \\ 1 & 2 \end{bmatrix} = \begin{bmatrix} -8-1 & 4-2 \\ 4+2 & -2+4 \end{bmatrix}$

$$= \begin{bmatrix} -9 & 2 \\ 6 & 2 \end{bmatrix}$$

45. a) $\begin{bmatrix} 1 & 3 \\ -1 & -2 \end{bmatrix}\begin{bmatrix} -2 & -3 \\ 1 & 1 \end{bmatrix} = \begin{bmatrix} -2+3 & -3+3 \\ 2-2 & +3-2 \end{bmatrix}$

$$= \begin{bmatrix} 1 & 0 \\ 0 & 1 \end{bmatrix}$$

   b) $\begin{bmatrix} -2 & -3 \\ 1 & 1 \end{bmatrix}\begin{bmatrix} 1 & 3 \\ -1 & -2 \end{bmatrix} = \begin{bmatrix} -2+3 & -6+6 \\ 1-1 & 3-2 \end{bmatrix}$

$$= \begin{bmatrix} 1 & 0 \\ 0 & 1 \end{bmatrix}$$

47. a) $\det[E] = 0$

   b) $[E]^{-1} = \begin{bmatrix} \dfrac{2}{0} & \dfrac{+1}{0} \\ \dfrac{4}{0} & \dfrac{2}{0} \end{bmatrix}$

   Zero denominators- no inverse
   is possible.
   See instructions for #27-45
   for inverse formula.

   c) Matrix with no inverse.
   (answers will vary with
   calculator)

49. $\det[A] = 0$
   The graphs are coincident.
   Three possible points are
   $(0,-4)$, $(1,-1)$, $(2,2)$

1. $5x - 2y = 4$

   $2y = 5x - 4$

   $y = \dfrac{5}{2}x - 2$

2. $2x - \dfrac{1}{2}y = 5$

   $\dfrac{1}{2}y = 2x - 5$

   $y = 4x - 10$

3. $2y - 3x = 8$

   $y = 1.5x - 3$

   $2(1.5x - 3) - 3x = 8$

   $3x - 6 - 3x = 8$

   $-6 = 8$

   false result - parallel lines
   No solution.

4. $3y = 2x - 6$

   $y = -\dfrac{5x}{6} + 1$

   $3\left(-\dfrac{5x}{6} + 1\right) = 2x - 6$

   $\dfrac{-5x}{2} + 3 = 2x - 6$

   $-5x + 6 = 4x - 12$

   $-9x = -18$

   $x = 2$

   $y = \dfrac{-5(2)}{6} + 1 = \dfrac{-5}{3} + \dfrac{3}{3} = \dfrac{-2}{3}$

   $\left(2, -\dfrac{2}{3}\right)$

5. $0.8x + y = -1.8$

   $1.2x + y = 3$

   $\begin{bmatrix} 0.8 & 1 \\ 1.2 & 1 \end{bmatrix}\begin{bmatrix} x \\ y \end{bmatrix} = \begin{bmatrix} -1.8 \\ 3 \end{bmatrix}$

   $\det[A] = -0.4$

   $\begin{bmatrix} x \\ y \end{bmatrix} = \begin{bmatrix} 12 \\ -11.4 \end{bmatrix}$

   $(12, -11.4)$

6. $-2.5x + y = 4$

   $-5x + 2y = 8$

   $\begin{bmatrix} -2.5 & 1 \\ -5 & 2 \end{bmatrix}\begin{bmatrix} x \\ y \end{bmatrix} = \begin{bmatrix} 4 \\ 8 \end{bmatrix}$

   $\det[A] = 0$

   $2(-2.5x + y) = 4(2)$

   $-5x + 2y = 8$

   Lines are coincident.

7. Answers will vary but
   $\det[A] = 0$.

8. $W + E = 54568$

   $W = E + 8270$

   $(E + 8270) + E = 54568$

   $2E = 46298$

   $E = 23149$

   $W = 23149 + 8270 = 31{,}419$

   Wilt 31,419;   Elgin 23,149

9. Slope at $(0, 600) = -0.05$

   $-0.05 = 2a(0) + b, \quad b = -0.05$

   $600 = a(0)^2 + (-0.05)(0) + c$

   $c = 600$

   at $(2000, 580)$ $slope = 0.03$

   $0.03 = 2a(2000) - 0.05$

   $0.08 = 4000a$

   $a = \dfrac{0.08}{4000} = \dfrac{8}{400{,}000} = \dfrac{1}{50{,}000}$

   $y = \dfrac{1}{50{,}000}x^2 - 0.05x + 600$

   $a = \dfrac{1}{50{,}000}, \quad b = -0.05, \quad c = 600$

10.

	Qty	Value	Qty · Value
water	x	0	0
peroxide	y	0.03	0.03y
	1 pint	0.02	0.02

   $0.03y = 0.02, \quad y = \dfrac{2}{3}$ pint

   $\dfrac{1}{3}$ pint water, $\dfrac{2}{3}$ pint peroxide

Section 7.3

1. $\begin{bmatrix} 1 & 1 & -1 \\ 1 & -1 & 1 \\ -1 & 1 & 1 \end{bmatrix} \begin{bmatrix} x \\ y \\ z \end{bmatrix} = \begin{bmatrix} 2 \\ 3 \\ 4 \end{bmatrix}$

$\begin{bmatrix} x \\ y \\ z \end{bmatrix} = \begin{bmatrix} 2.5 \\ 3 \\ 3.5 \end{bmatrix}$

$(2.5, 3, 3.5)$

3. $\begin{bmatrix} 0 & 3 & 5 \\ 3 & 0 & 7 \\ 1 & 1 & 4 \end{bmatrix} \begin{bmatrix} x \\ y \\ z \end{bmatrix} = \begin{bmatrix} 0 \\ 0 \\ 3 \end{bmatrix}$

$\det[A] = 0$

$3y + 5z = 0 \qquad y = -\dfrac{5}{3} z$

$3x + 7z = 0 \qquad x = -\dfrac{7}{3} z$

$\left(-\dfrac{7}{3} z \right) + \left(-\dfrac{5}{3} z \right) + 4z = 3$

$-\dfrac{12}{3} z + 4z = 3$

$-4z + 4z = 3$

$0 = 3$ False.

Equations are inconsistent.

5. $\begin{bmatrix} 2 & -1 & 0 \\ 1 & 0 & 2 \\ -2 & 1 & 1 \end{bmatrix} \begin{bmatrix} x \\ y \\ z \end{bmatrix} = \begin{bmatrix} 3 \\ 4 \\ 1 \end{bmatrix}$

$\begin{bmatrix} x \\ y \\ z \end{bmatrix} = \begin{bmatrix} -4 \\ -11 \\ 4 \end{bmatrix}$

$(-4, -11, 4)$

7. $\begin{bmatrix} -2 & 0 & 3 \\ -1 & 1 & 3 \\ 1 & 1 & 0 \end{bmatrix} \begin{bmatrix} x \\ y \\ z \end{bmatrix} = \begin{bmatrix} 0 \\ -1 \\ -1 \end{bmatrix}$

$\det[A] = 0$

$x + y = -1, \qquad y = -x - 1$

$-2x + 3z = 0, \qquad z = \dfrac{2}{3} x$

$-x + (-x - 1) + 3 \left(\dfrac{2}{3} x \right) = -1$

$-2x - 1 + 2x = -1$

$-1 = -1$ True

Equations are dependent.

9. $c + f + p = 32$

$c = f + 12 \qquad c - f = 12$

$4c + 9f + 4p = 148$

$\begin{bmatrix} 1 & 1 & 1 \\ 1 & -1 & 0 \\ 4 & 9 & 4 \end{bmatrix} \begin{bmatrix} c \\ f \\ p \end{bmatrix} = \begin{bmatrix} 32 \\ 12 \\ 148 \end{bmatrix}$

$\begin{bmatrix} c \\ f \\ p \end{bmatrix} = \begin{bmatrix} 16 \\ 4 \\ 12 \end{bmatrix}$

16 g Carbohydrates, 4 g fat
12 g protein

11. $n + d + q = 44$

$g = d + 3, \quad -d + q = 3$

$0.05n + 0.10d + 0.25q = 5.80$

$\begin{bmatrix} 1 & 1 & 1 \\ 0 & -1 & 1 \\ 0.05 & 0.10 & 0.25 \end{bmatrix} \begin{bmatrix} n \\ d \\ q \end{bmatrix} = \begin{bmatrix} 44 \\ 3 \\ 5.80 \end{bmatrix}$

$\begin{bmatrix} n \\ d \\ q \end{bmatrix} = \begin{bmatrix} 17 \\ 12 \\ 15 \end{bmatrix}$

17 nickels, 12 dimes 15 quarters

13. $A + S = P + 412{,}545$

$S = A + 459{,}907$

$A + P = 1{,}445{,}464$

$P = 1{,}445{,}464 - A$

$A + (A + 459{,}907)$

$= (1{,}445{,}464 - A) + 412{,}545$

$2A + 459{,}907 = 1{,}858{,}009 - A$

$3A = 1{,}398{,}102$

$A = 466{,}034$

$S = 466{,}034 + 459{,}907 = 925{,}941$

$466{,}034 + P = 1{,}445{,}464$

$P = 979{,}430$

(Ayako Okamoto) $466{,}034

(Curtis Strange) $925{,}941

(Corey Pavin) $979{,}430

15.

	Qty	Value	Qty·Value
Ethiopian	E	10.95	10.95E
French	F	8.95	8.95F
Guatemalan	G	8.50	8.50G
Colombian	C	8.25	8.25C
Total	1000	8.59	8590

$E + F + G + C = 1000$

$10.95E + 8.95F + 8.50G + 8.25C = 8590$

$C = 2G, \quad 2G - C = 0$

$F = 4E, \quad 4E - F = 0$

$$\begin{bmatrix} 1 & 1 & 1 & 1 \\ 10.95 & 8.95 & 8.5 & 8.25 \\ 0 & 0 & 2 & -1 \\ 4 & -1 & 0 & 0 \end{bmatrix} \begin{bmatrix} E \\ F \\ G \\ C \end{bmatrix} = \begin{bmatrix} 1000 \\ 8590 \\ 0 \\ 0 \end{bmatrix}$$

$$\begin{bmatrix} E \\ F \\ G \\ C \end{bmatrix} \cong \begin{bmatrix} 50 \\ 200 \\ 250 \\ 500 \end{bmatrix}$$

50 pounds Ethiopian,
200 pounds French Roast,
250 pounds Guatemalan,
500 pounds light Colombian

17.
$$\begin{bmatrix} (-2)^4 & (-2)^3 & (-2)^2 & (-2)^1 & 1 \\ (-1)^4 & (-1)^3 & (-1)^2 & (-1)^1 & 1 \\ 0 & 0 & 0 & 0 & 1 \\ 1 & 1 & 1 & 1 & 1 \\ 2^4 & 2^3 & 2^2 & 2 & 1 \end{bmatrix} \begin{bmatrix} a \\ b \\ c \\ d \\ e \end{bmatrix} = \begin{bmatrix} -7 \\ -9 \\ -5 \\ -1 \\ 45 \end{bmatrix}$$

$$\begin{bmatrix} a \\ b \\ c \\ d \\ e \end{bmatrix} = \begin{bmatrix} 2 \\ 3 \\ -2 \\ 1 \\ -5 \end{bmatrix}$$

$y = ax^4 + bx^3 + cx^2 + dx + e$

$y = 2x^4 + 3x^3 - 2x^2 + x - 5$

19.
$$\begin{bmatrix} 1 & 1 & 1 \\ 1 & -1 & 0 \\ 0 & 0 & 1 \end{bmatrix} \begin{bmatrix} s \\ m \\ 1 \end{bmatrix} = \begin{bmatrix} 620 \\ 0 \\ 20 \end{bmatrix}$$

21. a) $[A]$ is 1×3, $[B]$ is 3×4
   product $[A][B]$ is 1x4

   b) $[C] = \begin{bmatrix} 4800 & 5100 & 5400 & 6000 \end{bmatrix}$

23. a) Small $\dfrac{4.5}{16}$, medium $\dfrac{12}{16}$ or $\dfrac{3}{4}$, large 3.

   b) $\begin{bmatrix} \dfrac{4.5}{16} & \dfrac{12}{16} & 3 \end{bmatrix} \begin{bmatrix} 500 & 450 & 400 & 300 \\ 100 & 150 & 200 & 300 \\ 20 & 20 & 20 & 20 \end{bmatrix}$

   $\approx \begin{bmatrix} 275.6 & 299.1 & 322.5 & 369.4 \end{bmatrix}$

   c) $3.50 \cdot \begin{bmatrix} 275.6 & 299.1 & 322.5 & 369.4 \end{bmatrix}$

   $\approx \begin{bmatrix} \$965 & \$1047 & \$1129 & \$1293 \end{bmatrix}$

25. $[C] - [D] - [E] =$

   $\begin{bmatrix} 4800 & 5100 & 5400 & 6000 \end{bmatrix}$

   $- \begin{bmatrix} 522 & 507 & 492 & 462 \end{bmatrix}$

   $- \begin{bmatrix} 965 & 1047 & 1129 & 1293 \end{bmatrix}$

   $\approx \begin{bmatrix} \$3313 & \$3546 & \$3779 & \$4245 \end{bmatrix}$

Section 7.3

27. Because a system with no solution has a false statement within its solution, one method is to write equations whose variables add to zero while its constants add to a non-zero number. Example:

$$x + y + z = -3$$
$$x - y \quad\;\; = 4$$
$$\underline{-2x \quad - z = 5}$$
$$0 \quad 0 \quad 0 = 6$$

Section 7.4

1. Parabola;
   y-intercept = $(0, 2)$

3. Circle:
   y-intercepts $0^2 + y^2 = 9$
   $y = \pm 3$
   $(0, \pm 3)$;
   x – intercepts $= (\pm 3, 0)$;
   To graph in claculator
   Solve for y:
   $(y^2 = 9 - x^2)$
   $y = \pm\sqrt{9 - x^2}$

5. Hyperbola;
   x-intercepts $= (\pm 2, 0)$;
   To graph in calculator:
   $y^2 = x^2 - 4$
   $y = \pm\sqrt{x^2 - 4}$

7. Ellipse:
   x-intercepts
   $\frac{1}{4}x^2 + 0 = 1$
   $x^2 = 4, \quad x = \pm 2$
   $(\pm 2, 0)$;

   y – intercepts; $\frac{1}{4} \cdot 0 + y^2 = 1$
   $y = \pm 1$
   $(0, \pm 1)$;

   To graph in calculator:
   $y^2 = 1 - \frac{1}{4}x^2$
   $y = \pm\sqrt{1 - \frac{x^2}{4}}$

9. Ellipse;
   x-intercepts
   $4x^2 + 0 = 100$
   $x^2 = 25, \quad x = \pm 5$
   $(\pm 5, 0)$;

   y – intercepts; $4 \cdot 0 + y^2 = 100$
   $y = \pm 10$
   $(0, \pm 10)$;

   To graph in calculator:
   $y^2 = 100 - 4x^2 = 4(25 - x^2)$
   $y = \pm\sqrt{4(25 - x^2)} = \pm 2\sqrt{25 - x^2}$

11. Circle;
    x-intercepts
    $2x^2 + 2 \cdot 0 = 8$
    $x^2 = 4, \quad x = \pm 2$
    $(\pm 2, 0)$;

    y – intercepts $= (0, \pm 2)$;

    To graph in calculator:
    $2y^2 = 8 - 2x^2$
    $y^2 = 4 - x^2$
    $y = \pm\sqrt{4 - x^2}$

13. Straight line;
    x-and y-intercept $= (0, 0)$.

15. Hyperbola;
    x-intercepts; $x^2 - 4 \cdot 0 = 1$
    $x = \pm 1$
    $(\pm 1, 0)$;

    To graph in calculator:
    $4y^2 = x^2 - 1$
    $y^2 = \frac{1}{4}(x^2 - 1)$
    $y = \pm\sqrt{\frac{1}{4}(x^2 - 1)}$
    $y = \pm\frac{1}{2}\sqrt{x^2 - 1}$

17. Parabola;
    x and y-intercept = (0, 0);
    To graph in calculator:

    $4y^2 = x$

    $y^2 = \frac{1}{4}x$

    $y = \pm\sqrt{\frac{1}{4}x}$

    $y = \pm\frac{1}{2}\sqrt{x}$

19. Hyperbola;

    y-intercept; $y^2 - \frac{0}{4} = 1$

    $\qquad\qquad y = \pm 1$

    $\qquad\qquad (0, \pm 1)$;

    To graph in claculator:

    $y^2 = \frac{x^2}{4} + 1$

    $y = \pm\sqrt{\frac{x^2}{4} + 1}$.

21. Parabola;
    x and y intercept = (0,0)
    To graph in a calculator:

    $y^2 = -x$

    $y = \pm\sqrt{-x}$

23. $r = 4$, $\quad x^2 + y^2 = 4^2$

    $\qquad\qquad x^2 + y^2 = 16$

25. $\frac{x^2}{1^2} + \frac{y^2}{5^2} = 1$, $\quad \frac{x^2}{1} + \frac{y^2}{25} = 1$

27. $\frac{y^2}{3^2} - \frac{x^2}{5^2} = 1$, $\quad \frac{y^2}{9} - \frac{x^2}{25} = 1$

29. $1 = a(-2)^2$, $\quad y = \frac{1}{4}x^2$

    $a = \frac{1}{4}$

31. Negative length has no meaning. If r = 0 the circle is a point.

33. The branches become steeper as b^2 gets large.

35. Answers may vary, possible answers are: $(\pm 3, 0)$ $\quad (0, \pm 3)$

37. Answers may vary, possible answers are: $(\pm n, 0)$ $\quad (0, \pm n)$

39. Answers may vary, possible answers are:
    $(1, -1)$, $\quad (-5, -1)$, $\quad (-2, 2)$, $\quad (-2, -4)$

41. answers may vary, possible answers are:
    $(\pm 3, \pm 4)$, $\quad (\pm 4, \pm 3)$, $\quad (\pm 5, 0)$, $\quad (0, \pm 5)$

43. $d = \sqrt{\left((x - 0)^2 + (y - n)^2\right)}$

    $d = \sqrt{x^2 + (y - n)^2}$

45. The origin is the same distance, n, from the point F(0,n) as from the directrix line y = -n as required by definition. The v refers to vertex.

47. $\frac{1}{4n} = 1$, $\quad n = \frac{1}{4}$; parabola opens up. Focus at $\left(0, \frac{1}{4}\right)$.

49. $\frac{1}{4n} = 0.5$, $\quad 0.25 = 0.5n$, $\quad n = \frac{0.25}{0.5} = \frac{1}{2}$
    Parabola opens to right. Focus at $\left(\frac{1}{2}, 0\right)$.

51. $\frac{1}{4n} = -1$, $\quad n = -\frac{1}{4}$
    Parabola opens down. Focus at $\left(0, -\frac{1}{4}\right)$.

53. $d = r = \sqrt{(x - h)^2 + (y - k)^2}$

$r^2 = (x - h)^2 + (y - k)^2$

55. when $x \leq 0$.

Section 7.5

1. Answers will vary: Possible window:
   x min: -1.1 x max: 0.1
   y min: -1.1 y max: 0.1

3. $0 \overset{?}{=} -2(-1)^2 + 2$ $0 \overset{?}{=} -2(1)^2 + 2$
   $0 = -2 \cdot 1 + 2$ $0 = -2 \cdot 1 + 2$
   $0 = 0 \checkmark$ $0 = 0 \checkmark$

   $(-1)^2 + (0-2)^2 \overset{?}{=} 4$ $(1)^2 + (0-2)^2 \overset{?}{=} 4$
   $1 + 4 = 4$ $1 + 4 = 4$
   $5 \neq 4$ no $5 \neq 4$ no
   Not solutions to the system. Satisfy first equation but not the second.

5. $y = x^2 - 3x + 2 \rightarrow$ parabola
   $y = -x + 5 \rightarrow$ line
   $-x + 5 = x^2 - 3x + 2$
   $0 = x^2 - 2x - 3$
   $(x - 3)(x + 1) = 0$
   $x = 3, \quad x = -1$

   $y = -(3) + 5 = 2$
   $y = -(-1) + 5 = 6$

   $(-1, 6), (3, 2)$

7. $\dfrac{x^2}{4} - \dfrac{y^2}{9} = 1 \rightarrow$ hyperbola
   $y = x^2 - 4 \rightarrow$ parabola
   $x^2 = y + 4$
   $36\left(\dfrac{y+4}{4} - \dfrac{y^2}{9} = 1\right)$
   $9(y + 4) - 4y^2 = 36$
   $9y + 36 - 4y^2 = 36$
   $-4y^2 + 9y = 0$
   $y(-4y + 9) = 0$
   $y = 0, \quad y = \dfrac{9}{4} = 2.25$
   $0 = x^2 - 4, \quad x^2 = 4,$
   $x = \pm 2$
   $2.25 = x^2 - 4, \quad x^2 = 6.25,$
   $x = \pm 2.5 \quad (\pm 2, 0), \quad (\pm 2.5, 2.25)$

9. $y = 5x^2 + 2x \rightarrow$ parabola
   $y = -x^2 + x + 2 \rightarrow$ parabola
   $5x^2 + 2x = -x^2 + x + 2$
   $6x^2 + x - 2 = 0$
   $(3x + 2)(2x - 1) = 0$
   $x = -\dfrac{2}{3}, \quad x = \dfrac{1}{2}$

   $y = 5\left(-\dfrac{2}{3}\right)^2 + 2\left(-\dfrac{2}{3}\right)$
   $y = 5 \cdot \dfrac{4}{9} - \dfrac{4}{3} = \dfrac{20}{9} - \dfrac{12}{9} = \dfrac{8}{9}$

   $y = 5\left(\dfrac{1}{2}\right)^2 + 2\left(\dfrac{1}{2}\right)$
   $y = 5 \cdot \dfrac{1}{4} + 1 = \dfrac{5}{4} + \dfrac{4}{4} = \dfrac{9}{4}$

   $\left(-\dfrac{2}{3}, \dfrac{8}{9}\right), \quad \left(\dfrac{1}{2}, \dfrac{9}{4}\right)$

11. $x^2 + \dfrac{y^2}{4} = 1 \rightarrow$ ellipse

    $y = x^2 - 1 \rightarrow$ parabola

    $x^2 = y + 1$

    $y + 1 + \dfrac{y^2}{4} = 1$

    $4y + 4 + y^2 = 4$

    $y^2 + 4y = 0$

    $y(y + 4) = 0$

    $y = 0, \quad y = -4$

    $0 = x^2 - 1, \; x^2 = 1, \; x = \pm 1.$

    $-4 = x^2 - 1, \; x^2 = -3,$
    not possible, extraneous
    answer.
    $(1, 0), \quad (-1, 0)$

13. $x^2 + y^2 = 16 \rightarrow$ circle

    $y = x^2 - 4 \rightarrow$ parabola

    $x^2 = y + 4$

    $y + 4 + y^2 = 16$

    $y^2 + y - 12 = 0$

    $(y + 4)(y - 3) = 0$

    $y = -4, \; y = 3$

    $-4 = x^2 - 4, \; x^2 = 0, \; x = 0$

    $3 = x^2 - 4, \; x^2 = 7, \; x = \pm \sqrt{7}.$

    $\left(-\sqrt{7}, 3\right), \; \left(\sqrt{7}, 3\right)(0, -4)$

15. $y = -x^2 + 2 \rightarrow$ parabola

    $y = x^2 - 6 \rightarrow$ parabola

    $-x^2 + 2 = x^2 - 6$

    $0 = 2x^2 - 8$

    $2x^2 = 8$

    $x^2 = 4$

    $x = \pm 2$

    $y = (-2)^2 - 6 = 4 - 6 = -2$

    $y = 2^2 - 6 = 4 - 6 = -2$

    $(-2, -2), \; (2, -2)$

17. $\dfrac{x^2}{4} + y^2 = 1 \rightarrow$ ellipse

    $y = x^2 - 1 \rightarrow$ parabola

    $x^2 = y + 1$

    $\dfrac{y + 1}{4} + y^2 = 1$

    $y + 1 + 4y^2 = 4$

    $4y^2 + y - 3 = 0$

    $(4y - 3)(y + 1) = 0$

    $y = \dfrac{3}{4}, \; y = -1$

    $\dfrac{3}{4} = x^2 - 1$

    $x^2 = 1.75, \; x = \pm \sqrt{1.75} \approx \pm 1.323.$

    $-1 = x^2 - 1, \; x^2 = 0, \; x = 0$

    $(0, -1), \; \left(\pm \sqrt{1.75}, 0.75\right)$

Section 7.5

19. $x^2 + y^2 = 4 \rightarrow$ circle

    $x^2 - y^2 = 4 \rightarrow$ hyperbola

    $y^2 = x^2 - 4$

    $x^2 + x^2 - 4 = 4$

    $2x^2 = 8$

    $x^2 = 4$

    $x = \pm 2$

    $y^2 + 2^2 = 4,\ y^2 = 0,\ y = 0.$

    $y^2 + (-2)^2 = 4,\ y^2 = 0,\ y = 0.$

    $(-2, 0),\ (2, 0).$

21. $Y = 3^X,\ Y = 2 - 3^X$

    $3^X = 2 - 3^X$

    $2 \cdot 3^X = 2$

    $3^X = 1$

    $X = 0$

    $Y = 1$

    $(0, 1)$

23. $Y = 2^{X+2},\ Y = 4 + 2^X$

    $2^{X+2} = 4 + 2^X$

    $2^X \cdot 2^2 = 4 + 2^X$

    $4 \cdot 2^X = 4 + 2^X$

    $3 \cdot 2^X = 4$

    $2^X = \dfrac{4}{3}$

    $\log 2^x = \log \dfrac{4}{3}$

    $x \log 2 = \log \dfrac{4}{3}$

    $x = \dfrac{\log \dfrac{4}{3}}{\log 2} \approx 0.415$

    $y = 4 + 2^{0.415} \approx 5.333.$

    $(0.415,\ 5.333)$

25. $y = \log x,\quad y = 3 - \log x$

    $\log x = 3 - \log x$

    $2 \log x = 3$

    $\log x^2 = 3$

    $10^3 = x^2$

    $x = 10^{3/2} \approx 31.6$ (discard negative solution)

    $y = \log 31.6 \approx 1.5.$

    $(31.6, 1.5)$

27. $\log_2 x^2 = 1$

    $2^1 = x^2$

    $x = \pm \sqrt{2}$

    Discard negative solution because it is not in the domain of the logarithmic function. Answer is the same.

Chapter 7 Review

1. $\ell + w + h = 12$

   $\ell \cdot w = 15$

   $w \cdot h = 12$

   $3 \cdot 5 = 15$

   $3 + 5 + h = 12$

   $h = 4$

   $3 \cdot 4 = 12$

   One possible dimension set:
   5 in, 4 in, 3 in.

   $\ell = \dfrac{15}{w}, \; h = \dfrac{12}{w}$

   $\dfrac{15}{w} + w + \dfrac{12}{w} = 12$

   $15 + w^2 + 12 = 12w$

   $w^2 - 12w + 27 = 0$

   $(w - 3)(3 - 9) = 0$

   $w = 3$ or $w = 9$

   $\ell = \dfrac{15}{9} = 1\dfrac{2}{3}$

   $h = \dfrac{12}{9} = 1\dfrac{1}{3}$

   Other possible solution:

   9 in, $1\dfrac{2}{3}$ in, $1\dfrac{1}{3}$ in.

3. $\ell \cdot w \cdot h = 240$

   $\ell = w + h$

   $w = h + 2$

   $\ell = h + 2 + h = 2h + 2$

   $(2h + 2)(h + 2)h = 240$

   $2h^3 + 6h^2 + 4h - 240 = 0$
   From there you can solve
   graphically or with calculator.
   $h = 4$, $w = 4 + 2 = 6$,
   $\ell = 4 + 6 = 10$.

   $\ell = 10$ in, $w = 6$ in, $h = 4$ in.

5. $3x + 5y = 5$

   $2x - 4y = 18$

   $\begin{bmatrix} 3 & 5 \\ 2 & -4 \end{bmatrix} \begin{bmatrix} x \\ y \end{bmatrix} = \begin{bmatrix} 5 \\ 18 \end{bmatrix}$

   $\begin{bmatrix} x \\ y \end{bmatrix} = \begin{bmatrix} 5 \\ -2 \end{bmatrix}$

   $y = 5, \; y = -2$

7. $1.5x = 3.5y + 19, \quad 1.5x - 3.5y = 19$

   $5.5x + 1.5y = 41$

   $\begin{bmatrix} 1.5 & -3.5 \\ 5.5 & 1.5 \end{bmatrix} \begin{bmatrix} x \\ y \end{bmatrix} \begin{bmatrix} 19 \\ 41 \end{bmatrix}$

   $\begin{bmatrix} x \\ y \end{bmatrix} = \begin{bmatrix} 8 \\ -2 \end{bmatrix}$

   $x = 8, \; y = -2$

9. $-3x + 4y = 5$

   $y = \dfrac{3}{4}x + 2$

   $-3x + 4\left(\dfrac{3}{4}x + 2\right) = 5$

   $-3x + 3x + 8 = 5$
   $8 = 5$ false result;
   no solution, parallel lines.

11. $3x - y + z = 4$

    $-x \quad\;\; + z = 2, \quad x = z - 2$

    $-y + 4z = 10, \quad y = 4z - 10$

    $3(z - 2) - (4z - 10) + z = 4$

    $3z - 6 - 4z + 10 + z = 4$

    $4 = 4$

    Variable dropped out-true result,
    dependent system.

13. $2x + y - z = 3$

    $x + y + 2z = 4$

    $x - y - 3z = 6$

    $$\begin{bmatrix} 2 & 1 & -1 \\ 1 & 1 & 2 \\ 1 & -1 & -3 \end{bmatrix}\begin{bmatrix} x \\ y \\ z \end{bmatrix} = \begin{bmatrix} 3 \\ 4 \\ 6 \end{bmatrix} \qquad \begin{bmatrix} x \\ y \\ z \end{bmatrix} = \begin{bmatrix} 6.2 \\ -7 \\ 2.4 \end{bmatrix}$$

    $x = 6.2, \ y = -7, \ z = 2.4$

15. $y = x^2 + x - 2 \rightarrow$ parabola

    $y = -x - 3 \rightarrow$ line

    $x^2 + x - 2 = -x - 3$

    $x^2 + 2x + 1 = 0$

    $(x + 1)(x + 1) = 0$

    $x = -1$

    $y = -(-1) - 3 = 1 - 3 = -2$

    $(-1, -2)$

17. $y = x^2 - 4x - 1 \rightarrow$ parabola

    $y = -x + 3 \rightarrow$ line

    $x^2 - 4x - 1 = -x + 3$

    $x^2 - 3x - 4 = 0$

    $(x - 4)(x + 1) = 0$

    $x = 4, \ x = -1$

    $y = -(-1) + 3 = 1 + 3 = 4.$

    $y = -(-4) + 3 = -4 + 3 = -1.$

    $(-1, 4) , \ (4, -1)$

19. $x^2 + y^2 = 16 \rightarrow$ circle

    $y = x^2 + 1 \rightarrow$ parabola

    $x^2 = y - 1$

    $y - 1 + y^2 = 16$

    $y^2 + y - 17 = 0$

    $y \cong -4.653, \quad y \cong 3.653$

    $-4.653 = x^2 + 1, \ x^2 = -5.653,$ not possible - extraneous answer.

    $3.653 = x^2 + 1, \ x^2 = 2.653,$

    $x \cong \pm 1.629$

    $(\pm 1.629, 3.653) .$

21. $y = x^2 - x \rightarrow$ parabola

    $y = -x^2 + 3 \rightarrow$ parabola

    $x^2 - x = -x^2 + 3$

    $2x^2 - x - 3 = 0$

    $(2x - 3)(x + 1) = 0$

    $x = \dfrac{3}{2}, \quad x = -1$

    $y = -\left(\dfrac{3}{2}\right)^2 + 3 = -\dfrac{9}{4} + \dfrac{12}{4} = \dfrac{3}{4}$

    $y = -(-1)^2 + 3 = -1 + 3 = 2$

    $(-1, 2), \quad \left(\dfrac{3}{2}, \dfrac{3}{4}\right).$

23. $y = x^2 + 1 \rightarrow$ parabola

    $y = x^2 - 5 \rightarrow$ parabola

    $x^2 + 1 = x^2 - 5$

    $1 = -5$ False - no solution.

    They do not intersect.

25. $y = 2^x, \quad y = 4 - 2^x$

$2^x = 4 - 2^x$

$2 \cdot 2^x = 4$

$2^x = 2$

$x = 1$

$y = 2^1 = 2$

$x = 1, \quad y = 2$

27. $y = \log x, \quad y = 3 - 2 \log x$

$\log x = 3 - 2 \log x$

$\log x = 3 - \log x^2$

$\log x + \log x^2 = 3$

$\log x^3 = 3$

$10^3 = x^3$

$x = 10$

$y = \log 10$

$y = 1$

$x = 10, \quad y = 1$

29. $J = 2 + T$

$J + T = 40$

$2 + T + T = 40$

$2T = 38$

$T = 19$

$J + 19 = 40$

$J = 21$

Jeanne, 21; Tish, 19.

31. $c + p + i = 215$

$p = 2c - 1$

$p = 47i$

$c + p + i = 215$

$2c - p \quad = 1$

$p - 47i = 0$

$$\begin{bmatrix} 1 & 1 & 1 \\ 2 & -1 & 0 \\ 0 & 1 & -47 \end{bmatrix} \begin{bmatrix} c \\ p \\ i \end{bmatrix} = \begin{bmatrix} 215 \\ 1 \\ 0 \end{bmatrix}$$

$$\begin{bmatrix} c \\ p \\ i \end{bmatrix} = \begin{bmatrix} 71 \\ 141 \\ 3 \end{bmatrix}$$

71 mg calcium,   141 mg phosphorus,
3 mg iron.

33. $p + f + c = 26$

$f = p + 9, \quad -p + f = 9$

$4p + 9f + 4c = 184$

$$\begin{bmatrix} 1 & 1 & 1 \\ -1 & 1 & 0 \\ 4 & 9 & 4 \end{bmatrix} \begin{bmatrix} p \\ f \\ c \end{bmatrix} = \begin{bmatrix} 26 \\ 9 \\ 184 \end{bmatrix}$$

$$\begin{bmatrix} p \\ f \\ c \end{bmatrix} = \begin{bmatrix} 7 \\ 16 \\ 3 \end{bmatrix}$$

7g protein, 16g fat,
3g carbohydrate.

35. $10T + 5F + 1 \cdot N = 86$

$F = T + N - 2, \quad T - F + N = 2$

$T + F + N = 18$

$$\begin{bmatrix} 10 & 5 & 1 \\ 1 & -1 & 1 \\ 1 & 1 & 1 \end{bmatrix} \begin{bmatrix} T \\ F \\ N \end{bmatrix} = \begin{bmatrix} 86 \\ 2 \\ 18 \end{bmatrix}$$

$$\begin{bmatrix} T \\ F \\ N \end{bmatrix} = \begin{bmatrix} 4 \\ 8 \\ 6 \end{bmatrix}$$

4 tens, 8 fives, 6 ones.

37. 2800ft to the left of $(2800, 300)$

$x = 0$

$slope, m = -0.03 = 2a(0) + b,$

$b = -0.03 = -\dfrac{3}{100}.$

Slope at $(2800, 300) = 0.04$

$0.04 = 2a(2800) - 0.03$

$0.07 = 5600a$

$a = \dfrac{0.07}{5600} = \dfrac{7}{560,000} = \dfrac{1}{80,000}.$

$300 = \dfrac{1}{80,000}(2800)^2 - \dfrac{3}{100}(2800) + c$

$300 = 98 - 84 + c$

$286 = c$

$a = \dfrac{1}{80,000}, \ b = -\dfrac{3}{100}, \ c = 286.$

Chapter 7 Test

1. $\ell \cdot w \cdot h = 240$
    $\ell \cdot w = 20$
    $\ell + w + h = 24$
    $20 \cdot h = 240$
    $h = 12$
    $\ell + w + 12 = 24$
    $\ell + w = 12$
    $\ell \cdot w = 20$
    $2 \cdot 10 = 20,\ 2 + 10 = 12.$
    Dimensions are:
    2in., 10in, 12in.

2. $2x - 3y = 6$
    $4x + 5y = -32$
    $\begin{bmatrix} 2 & -3 \\ 4 & 5 \end{bmatrix} \begin{bmatrix} x \\ y \end{bmatrix} = \begin{bmatrix} 6 \\ -32 \end{bmatrix}$
    $\begin{bmatrix} x \\ y \end{bmatrix} = \begin{bmatrix} -3 \\ -4 \end{bmatrix}$
    $x = -3,\ y = -4$

3. $x + 2y = -8, \quad x = -8 - 2y$
    $0.5x + y = -4$
    $0.5(-8 - 2y) + y = -4$
    $-4 - y + y = -4$
    $-4 = -4;$ True result,
    coincident lines.

4. $2x + y + 2z = 5$
    $x + 3z = 4, \quad z = -\dfrac{1}{3}x + \dfrac{4}{3}$
    $4x + 3y = 0, \quad y = -\dfrac{4}{3}x$
    $2x + -\dfrac{4}{3}x + 2\left(-\dfrac{1}{3}x + \dfrac{4}{3}\right) = 5$
    $2x + -\dfrac{4}{3}x + \dfrac{-2}{3}x + \dfrac{8}{3} = 5$
    $6x - 4x - 2x + 8 = 15$
    $8 = 15$
    False result, no solution.
    Inconsistent.

5. $3x + y + z = 8$
    $2x - y - 2z = 17$
    $-x - y + 3z = -9$
    $\begin{bmatrix} 3 & 1 & 1 \\ 2 & -1 & -2 \\ -1 & -1 & 3 \end{bmatrix} \begin{bmatrix} x \\ y \\ z \end{bmatrix} = \begin{bmatrix} 8 \\ 17 \\ -9 \end{bmatrix}$
    $\begin{bmatrix} x \\ y \\ z \end{bmatrix} = \begin{bmatrix} 4.5 \\ -3 \\ -2.5 \end{bmatrix}$
    $x = 4.5, \quad y = -3, \quad z = -2.5$

6. $y = 3x^2 - 4x - 5 \rightarrow$ parabola
    $y = x - 3 \rightarrow$ line
    $3x^2 - 4x - 5 = x - 3$
    $3x^2 - 5x - 2 = 0$
    $(3x + 1)(x - 2) = 0$
    $x = -\dfrac{1}{3}, \quad x = 2$
    $y = -\dfrac{1}{3} - 3 = -\dfrac{10}{3}$
    $y = 2 - 3 = -1$
    $\left(-\dfrac{1}{3}, -\dfrac{10}{3}\right), \quad (2, -1).$

7. $y = x^2 - 3x + 2 \rightarrow$ parabola
    $y = \dfrac{1}{2}x - 2 \rightarrow$ line
    $x^2 - 3x + 2 = \dfrac{1}{2}x - 2$
    $x^2 - 3.5x + 4 = 0$
    $\sqrt{(-3.5)^2 - 4 \cdot 1 \cdot 4} = \sqrt{12.25 - 16}$
    $= \sqrt{-3.75}$
    No real number solution,
    no point of intersection.

Chapter 7 Test

8. $x^2 + y^2 = 4 \rightarrow$ circle

$-x^2 + y^2 = 4 \rightarrow$ hyperbola, $y^2 = 4 + x^2$

$x^2 + 4 + x^2 = 4$

$2x^2 = 0, \quad x^2 = 0, \quad x = 0$

$0 + y^2 = 4, \quad y^2 = \pm 2$

$(0, \pm 2)$

9. $y = x^2 + 1 \rightarrow$ parabola, $x^2 = y - 1$

$x^2 + y^2 = 4 \rightarrow$ circle

$y - 1 + y^2 = 4$

$y^2 + y - 5 = 0$

$y \approx -2.791, \quad y \approx 1.791$

$-2.791 = x^2 + 1$

$x^2 = -3.791$

Not possible, extraneous solution.

$1.791 = x^2 + 1$

$x^2 = 0.791 = \pm\sqrt{0.791} \approx \pm 0.890$

$(\pm 0.890, 1.791)$

10. $y = 3^x, \quad y = 5 - 3^x$

$3^x = 5 - 3^x$

$2 \cdot 3^x = 5$

$3^x = \dfrac{5}{2}$

$\log_3 \dfrac{5}{2} = x$

$x = \dfrac{\log \dfrac{5}{2}}{\log 3} \approx 0.834.$

$y = 3^{0.834} = 2.5$

$(0.834, 2.5)$

11. $y = \log_3 x, \quad y = 4 - \log_3 x$

$\log_3 x = 4 - \log_3 x$

$\log_3 x + \log_3 x = 4$

$\log_3 x^2 = 4$

$3^4 = x^2$

$x = 3^2 = 9$

$y = \log_3 9$

$3^y = 9$

$3^y = 3^2$

$y = 2$

$(9, 2)$

12. $P = C + 10$

$P + C = 3922$

$C + 10 + C = 3922$

$2C = 3912$

$C = 1956$

Carol, 1956; Peggy, 1966.

13. $p = c - 3, \quad p - c = -3$

$p + c + f = 26$

$4p + 4c + 9f = 199$

$\begin{bmatrix} 1 & -1 & 0 \\ 1 & 1 & 1 \\ 4 & 4 & 9 \end{bmatrix} \begin{bmatrix} p \\ c \\ f \end{bmatrix} = \begin{bmatrix} -3 \\ 26 \\ 199 \end{bmatrix}$

$\begin{bmatrix} p \\ c \\ f \end{bmatrix} = \begin{bmatrix} 2 \\ 5 \\ 19 \end{bmatrix}$

2g protein, 5g carbohydrate, 19g fat.

14. $x + y = 400$

$0x + 0.16y = 0.014(400)$

$0.16y = 5.6$

$y = 35$

$x + 35 = 400, \quad x = 365$

365 gal water, 35 gal polyvinyl soln.

15. $p + n + q = 59$

$q = n + 10, \quad -n + q = 10$

$0.01p + 0.05n + 0.25q = 6.35$

$$\begin{bmatrix} 1 & 1 & 1 \\ 0 & -1 & 1 \\ 0.01 & 0.05 & 0.25 \end{bmatrix} \begin{bmatrix} p \\ n \\ q \end{bmatrix} = \begin{bmatrix} 59 \\ 10 \\ 6.35 \end{bmatrix}$$

$$\begin{bmatrix} p \\ n \\ q \end{bmatrix} = \begin{bmatrix} 25 \\ 12 \\ 22 \end{bmatrix}$$

25 pennies, 12 nickles,
22 quarters.

16. $0.89j + 0.59a + 3.98c = 15.07$

$j + a + c = 11$

$17j + 8a + 16c = 158$

$$\begin{bmatrix} 0.89 & 0.59 & 3.98 \\ 1 & 1 & 1 \\ 17 & 8 & 16 \end{bmatrix} \begin{bmatrix} j \\ a \\ c \end{bmatrix} = \begin{bmatrix} 15.07 \\ 11 \\ 158 \end{bmatrix}$$

$$\begin{bmatrix} j \\ a \\ c \end{bmatrix} = \begin{bmatrix} 6 \\ 3 \\ 2 \end{bmatrix}$$

6 juice, 3 apples, 2 Oreo's.

17. $m = -0.03 = 2a(0) + b$

$b = -0.03 = -\dfrac{3}{100}$

$m = 0.06 = 2a(3200) + (-0.03)$

$0.09 = 6400a$

$\dfrac{0.09}{6400} = a$

$a = \dfrac{9}{640,000}$

$1500 = \dfrac{9}{640,000}(0)^2 - \dfrac{3}{100}(0) + c$

$c = 1500$

$a = \dfrac{9}{640,000}, \quad b = \dfrac{-3}{100}, \quad c = 1500$

Jan 24 - MP

CC Bookstore
325 COLLEGE DRIVE
CASPER, WY 82601
307 265 2211

010000 9760534956684
INTERMEDIATE ALGEBRA

SA   GMP   905 08.46
YOUR SALES ASSOCIATE

****

THANK YOU FOR SHOPPING

****

PLEASE SAVE YOUR RECEIPT